The Oxford Poetry Library

GENERAL EDITOR: FRANK KERMODE

PHILIP SIDNEY was born on 30 November 1554. He was educated at Shrewsbury School and Christ Church, Oxford; he may have spent some time also at Cambridge. In 1572 he set out for three years of foreign travel. He was in Paris at the time of the St Bartholomew's Day Massacre of Protestants; he also stayed in Frankfurt, Vienna, Padua, and Venice (where Veronese painted his portrait, now lost), and made a brief excursion into Hungary. In 1577 he went to the Continent again, as an ambassador to the Imperial Court at Prague. Soon after his return he may have begun to write the *Old Arcadia*, which he finished in 1580. *The Defence of Poesy* probably belongs to 1581, and the sonnet sequence *Astrophil and Stella* to 1581–3. Soon after this he began to recast his *Arcadia* on epic lines, a revision which he never completed. In 1583 he married Frances Walsingham, whose father was Secretary of State to the Queen. Though he was appointed royal cup-bearer in 1576, and in 1583 was knighted, he held no major office until in 1585 he was appointed Governor of the Dutch port of Flushing. In September 1586 he was shot in the thigh during a battle against the Spanish near Zutphen and died of infection just over three weeks later.

KATHERINE DUNCAN-JONES was educated at King Edward VI High School for Girls, Birmingham, and St Hilda's College, Oxford, where she obtained the Charles Oldham Shakespeare Prize and the Matthew Arnold Memorial Prize. After a year of teaching in Cambridge she was elected to a fellowship at Somerville College, Oxford. In 1972 she edited (with J. van Dorsten) *Miscellaneous Prose of Sir Philip Sidney* for the Clarendon Press; her paperback *Selected Poems of Sir Philip Sidney* (also Clarendon Press) was published in 1973. She delivered the 1980 Chatterton Lecture to the British Academy on Sidney's poetry. Her biography *Sir Philip Sidney: Courtier Poet* was published in 1991. A longer selection of Sidney's work, also edited by Katherine Duncan-Jones, is available in The Oxford Authors series.

FRANK KERMODE, retired King Edward VII Professor of English Literature at Cambridge, is the author of many books, including *Romantic Image, The Sense of an Ending, The Classic, The Genius of Secrecy, Forms of Attention,* and *History and Value*; he is also co-editor with John Hollander of *The Oxford Anthology of English Literature.*

THE OXFORD POETRY LIBRARY

GENERAL EDITOR: FRANK KERMODE

Matthew Arnold	*Miriam Allott*
William Blake	*Michael Mason*
Byron	*Jerome McGann*
Samuel Taylor Coleridge	*Heather Jackson*
John Dryden	*Keith Walker*
Thomas Hardy	*Samuel Hynes*
George Herbert	*Louis Martz*
Gerald Manley Hopkins	*Catherine Phillips*
Ben Jonson	*Ian Donaldson*
John Keats	*Elizabeth Cook*
Andrew Marvell	*Frank Kermode and Keith Walker*
John Milton	*Jonathan Goldberg and Stephen Orgel*
Alexander Pope	*Pat Rogers*
Sir Philip Sidney	*Katherine Duncan-Jones*
Henry Vaughan	*Louis Martz*
William Wordsworth	*Stephen Gill and Duncan Wu*

The Oxford Poetry Library

Sir Philip Sidney

Edited by
KATHERINE DUNCAN-JONES

Oxford New York
OXFORD UNIVERSITY PRESS
1994

Oxford University Press, Walton Street, Oxford OX2 6DP
Oxford New York Toronto
Delhi Bombay Calcutta Madras Karachi
Kuala Lumpur Singapore Hong Kong Tokyo
Nairobi Dar es Salaam Cape Town
Melbourne Auckland Madrid
and associated companies in
Berlin Ibadan

Oxford is a trade mark of Oxford University Press

This selection first published in The Oxford Poetry Library 1994

British Library Cataloguing in Publication Data
Data available

Library of Congress Cataloging in Publication Data
Sidney, Philip, Sir, 1554–1586 [Selections. 1994]
Sir Philip Sidney / edited by Katherine Duncan-Jones.
p. cm. — (The Oxford poetry library)
Includes bibliographical references (p.).
I. Duncan-Jones, Katherine. II. Title. III. Series.
PR2341.D8 1994 821'.3—dc20 93-28759
ISBN 0-19-282274-8

1 3 5 7 9 10 8 6 4 2

Typeset by J&L Composition Ltd., Filey, North Yorkshire
Printed in Great Britain by
Biddles Ltd
Guildford and King's Lynn

Contents

Introduction

Whether we approach Sidney primarily through his literary works or his biography, the fact of his soldier's death in 1586, at the early age of 31, stops us short, casting long shadows back over all the rest. Yet the fact of his early death, so obvious to us, was not available to Sidney himself. Even within hours of his death he wrote a desperate letter to a doctor who might, he hoped, succeed in saving him though many others had failed.[1] More life was what he wanted, not the death of septicaemia that he endured. If Sidney's works are suffused with melancholy and terminate in sudden, inconclusive endings, the reasons must be sought elsewhere. When Sidney left England for the Netherlands in November 1585 he entrusted the manuscript of his revised *Arcadia* to his friend Fulke Greville, and presumably intended to resume his literary work in progress at a future date. The half-sentence with which the revised portion ends may have been a deliberate device to aid his memory when he got back to work on it. This ambitious, wide-ranging, and boldly exploratory romance marks the point Sidney had reached when he left England for active political and military service. The long 'New' *Arcadia* fragment leaves him, as a writer, still on the middle ground; still in process of self-discovery; still, perhaps, in quest of both the genre and the audience which might bring his powers to their greatest fulfilment.

The question of audience—for whom did Sidney write?—is an important one to ask, but a hard one to answer. It has a bearing on a yet more fundamental question: why did Sidney write at all? There is no evidence that he intended any of his works except his intemperate *Defence* of his uncle, the Earl of Leicester, to see print: it is to the accident of his death that we owe the fact of their early publication (see Chronology). Sidney's nearest contemporaries, writers such as Gascoigne, Spenser, and Lyly, wrote and published poetry in the hope of attracting or consolidating patronage, or of gaining secure employment and/or reputation at Court; Sidney did not need to make such connections. He had plenty of 'business', as he often complained in letters to his friends; and if writing poetry

[1] See K. Duncan-Jones, *Sir Philip Sidney: Courtier Poet* (1991), 302–3.

and fiction was one of the preoccupations that kept him away from Court for longish spells, it may even have hindered his chances of preferment. While other Elizabethans, Shakespeare included, wrote at least partly for money, Sidney may even have lost money by writing poetry, in so far as absorption in literary projects kept him out of the Queen's eye when jobs were being filled. A sense of career opportunities lost filters into several of the sonnets of *Astrophil and Stella*, such as 18:

> . . . my wealth I have most idly spent.
> My youth doth waste, my knowledge brings forth toys,
> My wit doth strive those passions to defend
> Which for reward spoil it with vain annoys.
> I see my course to lose itself doth bend . . .

As a young courtier, rather than a poet, Sidney appeared to have everything to play for. Unlike Spenser, who had to carve out a career for himself through diligent and loyal service to his employers, Sidney was from birth beset by more 'great expectation' (*AS* 21.8) than can have been altogether comfortable. His godfather was Philip II of Spain (then in England as the husband of Mary Tudor); Lady Jane Grey had been his aunt. As he grew up into a period in which many courtiers of the older generation were either unmarried, like Christopher Hatton and Edward Dyer, or, like Leicester, Cecil, and Walsingham, had difficulty in producing male heirs, he became a kind of universal 'nephew' figure. He was indeed nephew to two powerful noblemen, the Earls of Warwick and Leicester; and he was to become son-in-law and heir to the Queen's Principal Secretary, Sir Francis Walsingham. He filled similar roles abroad. William of Orange, leader of resistance to Spanish rule in the Netherlands, hoped that Sidney might marry one of his daughters; and he inspired more than fatherly devotion in the bachelor diplomat Hubert Languet. It is not clear that any of these avuncular figures were aware that the promising young courtier wrote poems and stories, or that they would have thought the better of him if they had. In imagination, at least, Sidney's 'heir apparent' role extended yet further. Reaching adulthood in the period of Queen Elizabeth's advancing middle age, with the last hopes of her producing either a consort or an heir to the throne collapsing at the end of the second Alençon courtship in 1582, Sidney became, in the eyes of some, almost a crown prince. During much of his life his father, Sir Henry Sidney, was Lord Deputy

Governor of Ireland, which came out in Latin as 'Pro-Rex'. As the son of a quasi-king he could be viewed, and on the Continent often was, as a quasi-prince. In addition, he had acquired a French barony in Paris when he was only 17. However, few of Sidney's many humanist friends realized that the young 'baron Sidney' was also a poet.

Why, then, did Sidney write poetry? In *A Defence of Poesy* he describes as having, in his 'not old years', 'slipped into the title of a poet' (101), and there may be some truth in this. He hints elsewhere in the *Defence*, and suggests more explicitly both in the 'Old' *Arcadia* and in *Astrophil and Stella*, that it was the experience of falling passionately in love that compelled Sidney/Philisides/Astrophil to relieve his pent-up anguish in verse. There may be some truth in this, too: we shall probably never know. But whether love or other factors were the chief catalyst, there may have been something accidental about Sidney's initial self-discovery as a poet. Oddly enough, given the restricted nature of his comments on drama in *A Defence of Poesy*, it probably had to do with drama. Both his school, Shrewsbury, and his Oxford college, Christ Church, used drama and performance as a regular educational activity. After his return from three years of European travel in 1575 Sidney played an increasingly public role in Elizabethan high society as a deviser of courtly entertainments. These included tiltyard appearances, playlets, and allegorical displays of varying degrees of sophistication. This public part of his literary output is represented here by the attractive pastoral mini-drama *The Lady of May*, enacted for the Queen in 1578 or 1579. Sidney's awareness of his fame and prestige as a participant in tournaments is reflected in *Astrophil and Stella*, especially in sonnets 41 and 53:

> In martial sports I had my cunning tried,
> And yet to break more staves did me address,
> While with the people's shouts, I must confess,
> Youth, luck and praise even filled my veins with pride . . . (53.1–4)

Perhaps it was through his participation in public entertainments of various kinds that Sidney came to discover his own extraordinary creative facility and the emotional release which it offered. Although in the *Defence of Poesy* he seems to align himself with the Puritan Stephen Gosson in his condemnation of the licentiousness of the public theatres—the 'abuse' of poetry—a feeling for drama and for theatrical effect characterizes many of his own works. Indeed, the earliest critical comment on *Astrophil and Stella*, Thomas Nashe's preface to the pirated edition of 1591, praises it as a 'theatre of pleasure': 'here you shall find a paper stage strewed with pearl, an artificial heaven to overshadow the fair frame, and crystal walls to encounter your curious eyes while the tragicomedy of love is performed by starlight'.[2] But for modern readers, if *Astrophil and Stella* is a drama, it is a psychodrama. The central relationship is not so much that between Astrophil and Stella as between Astrophil and himself. The title is unlikely to be Sidney's own; it could be aptly emended to *Astrophil and Astrophil*.

Sidney's most powerful and characteristic writing has little to do with the public displays from which it may have sprung. Early requirements to participate in social performances—learned, sporting, diplomatic, or courtly—may have enabled Sidney to discover his remarkable expressive skills. But he soon developed them in other directions, and for far more restricted audiences. It is clear that the *Arcadia*, in its original form, was for family entertainment only. As he said to his sister, 'now it is done only for you, only to you'. The audience he envisaged for *Astrophil and Stella* may have been even smaller. We do not know to what extent the sonnet sequence was composed for the immediate perusal of Stella's real-life model, Penelope Rich, née Devereux. She would certainly have been capable of appreciating it. Yet most often the 'felt' audience within individual sonnets is intensely private. If this is a 'theatre of pleasure', it often seems to have room only for an audience of one, Astrophil himself. Astrophil appears alienated both from his friends and from Stella as he hammers out his obsessions in neurotic solitude—'As good to write, as for to lie and groan' (40.1). In his *Defence of Poesy*, written during the same period of his life, 1581–3 Sidney rests his case for literature's value very firmly on its value as an adjunct to public life and public action. But his sonnet-lover Astrophil has remarkably little to say about civic virtue, and certainly does not exemplify it.

To one of the earliest and best of Sidney's modern critics it seemed that he wrote too much about love, and that his verse was not sufficiently varied. Yet he saw the power and originality of Sidney's lyric gift as inseparable from his obsessive emotionality:

> Once the poet has set himself the task of writing an amorous complaint, that deep melancholy which lay beneath the surface of glamour of

[2] Thomas Nashe, *Works*, ed. R. B. McKerrow (1966), iii. 329.

Elizabethan existence, and which was so characteristic of Sidney himself, begins to fill the conventional form with a more than conventional weight. It surges through the magical adagio of the lines; they have that depth of reverberation, like the sound of gongs beaten under water, which is sometimes characteristic of Sidney as of no other Elizabethan, not even Shakespeare.[3]

This is an eloquent account of the paradox at the heart of Sidney's writing. Not only is there a split between his theory and his poetic practice—the one public and moral, the other introverted and self-tormenting—but a contradiction is inherent in his very stylishness. The splendid rhetorical show of the public courtier, both in prose and verse, conceals inward torment. To a superficial view, the concealment may be effective. As Sidney said bitterly of the Queen, 'so long as she sees a silk doublet upon me her Highness will think me in good case'. Deep misery, time and again, wells up beneath the bright, witty, metrically assured surface of his verse lines. In the *Defence of Poesy* Sidney celebrates artistic freedom and joyous creativity. The unfettered imagination of the poet delights in 'ranging only within the zodiac of his own wit' as he delivers the 'golden world' of art (105). But this is not how it feels in many of his own lyrics and fictions, which present us with repeated images of paralysis and stagnation. In *Astrophil and Stella* the poet encloses emotional turbulence within an ordered framework, and makes his speaker astonishingly articulate about the impossibility of articulating true feeling, or articulating feeling truly. The true subject seems often to be not so much Astrophil's adoration of Stella as his inability to come to terms with his own inner self. Love, in Sidney's poetry, is intimately connected with depression and self-hatred, as in the brilliantly self-infolded 'double sestina' of Strephon and Klaius:

> I wish no evenings more to see, each evening;
> Shamed, I hate myself in sight of mountains,
> And stop mine ears lest I grow mad with music. (ll. 58–60)

Throughout his all too brief career, Sidney tended to write himself into corners. *Astrophil and Stella*, with its closing image of 'iron doors' (108.11) is finished rather than concluded. The final mood seems to be defined in the Eighth Song:

[3] Theodore Spencer, 'The Poetry of Sir Philip Sidney', *English Literary History*, 12 (1945), 267.

That therewith my song is broken.

The curses on those who despise poetry at the end of the *Defence* are so over the top as to make the reader wonder whether the whole treatise may not have been more than half a joke. Sidney's last work, the 'New' *Arcadia*, breaks off with three of its five principal characters (Pamela, Philoclea, Pyrocles) literally in prison, and the other two (Musidorus and Amphialus) trapped in misguided and violent conflict. John Carey has written powerfully of the 'constant impulse towards deadlock in the rhetoric' in the 'New' *Arcadia*, and of Sidney's accounts of 'the space in which the characters move, full of vacillation and ambivalence'.[4] Cruelty, misunderstanding, corruption, and siege warfare are themes which seem to prefigure situations that Sidney was soon to encounter in real life in the Netherlands in 1585–6, except that in the 'New' *Arcadia* it is sexual passion that is the central source of all these things.

If Sidney had lived longer, some external developments might have lifted his spirits. In particular, he would soon have found the confidence he expressed in the literary resources of the English language and English culture marvellously borne out. New developments would surely have stimulated fresh departures in his own later work. But he would have been surprised to discover that it was in drama, and above all in the public theatres, that the most exciting things were happening. His critical assumptions about drama in *A Defence of Poesy*, bound by an insistence on the classical unities, seem disappointingly restricted. But we should remember that his own early death was not the only event on the horizon that Sidney could not foresee. The *Defence* was probably written in 1581, when Shakespeare was only 17. Among others who were to light up the literary landscape in the later years of Elizabeth's reign were Marlowe, who was also 17; Nashe, who was 14; Donne, who was 9; and Ben Jonson, who was 8. If Sidney had lived to be 60 he could have seen all of Shakespeare's plays. Dying, as he did, at 31, he saw none.

[4] John Carey, 'Structure and Rhetoric in Sidney's *Arcadia*', in Dennis Kay (ed.), *Sir Philip Sidney: An Anthology of Modern Criticism* (1987), 246–64.

Chronology

1554 Sidney born at Penshurst, Kent, 30 November, eldest child of Sir Henry Sidney and Lady Mary, sister of Robert Dudley, Earl of Leicester. Philip II, who had married Mary Tudor in July, was his godfather.

1564 Enrolled at Shrewsbury School 17 October, along with his friend and future biographer Fulke Greville (1554–1628).

1565 Sir Henry Sidney began the first of three terms of office as Lord Deputy Governor of Ireland.

1566 Visited Kenilworth and Oxford, where he witnessed splendid entertainments for the Queen (August and September).

1567/8 Became an undergraduate at Christ Church, Oxford, where his contemporaries included William Camden, George Peele, Richard Hakluyt, Walter Ralegh, Thomas Bodley, Richard Hooker, and perhaps John Lyly.

1572 Began European tour as part of the train of the Earl of Lincoln (May); witnessed the massacre of Protestants in Paris, 24 August (St Bartholomew's Day); met the Protestant diplomat Hubert Languet in Strasbourg; spent winter in Frankfurt.

1573 Travelled to Venice by way of Heidelberg and the imperial court at Vienna.

1574 Studied in Padua and Venice, where his portrait (now lost) was painted by Paolo Veronese (February); visited Genoa and Florence (March); at Vienna with Hubert Languet, apart from an excursion to Poland in October (August–January).

1575 Returned to England by way of Brno, Prague, Dresden, Frankfurt, Heidelberg, and Antwerp, reaching London in June. With his parents and sister accompanied the Queen on progress to Kenilworth, Lichfield, Chartley, Stafford, Chillington, Dudley, and Worcester. Accompanied his father to Shrewsbury; returned to Court (August).

1576 Joined his father in Ireland (August–October), possibly travelling home with the body of Walter Devereux, 1st Earl of Essex, whose dying wish had been that his daughter Penelope ('Stella') should marry Philip. Succeeded his father as 'Cup-bearer' to the Queen.

1577 Sent as ambassador to the newly acceded Emperor Rudolph in Prague (February–June), renewing European contacts *en route* and meeting Don John of Austria, Edmund Campion, and (May) William of Orange. Visited Wilton, Wiltshire, where his sister Mary was newly married to the 2nd Earl of Pembroke (August–September); here he may have written 'A dialogue between two shepherds' and may have begun the 'Old' *Arcadia*. Wrote a defence of his father's policies in Ireland (*Misc. Prose*, 8–12). Participated in Accession Day Tilt, 17 November, and wrote songs for it.

1578 Possibly wrote *The Lady of May* for the Queen's visit to the Earl of Leicester at Wanstead (May); possibly accompanied the Queen on her summer progress, which included a visit to Audley End where she was entertained by representatives of the University of Cambridge.

1579 Visit to England by Hubert Languet and the German Protestant prince John Casimir (February); the latter was made a Knight of the Garter by the Queen. Quarrelled with the Earl of Oxford while playing tennis at Court (August); this dispute may have been connected with Sidney's opposition to the Queen's projected marriage to the Duke of Alençon, against which he argued at length in a widely circulated *Letter (Misc. Prose*, 33–57). Spenser dedicated *The Shepheardes Calender* to him (December).

1580 Probably completed the 'Old' *Arcadia*.

1581 Gave the Queen New Year's gift of 'a whip garnished with small diamonds'. Divided his time between London and Wilton. Became a Member of Parliament (April). Devised and participated in *The Triumph of the Four Foster Children of Desire* (May). Penelope Devereux married Lord Rich (1 November). Sidney perhaps wrote 'Lamon's Tale' and began to write *Astrophil and Stella* and *The Defence of Poesy* about this time.

1582 Among large party of nobility and gentry which accompanied the Duke of Alençon to Antwerp (February); took an increasingly active interest in schemes to colonize America.

1583 Stood proxy for the absent Prince Casimir at his official installation as Knight of the Garter at Windsor Castle (see above, 1579), and as a result, became a knight (January). Married Frances, daughter and heir of the Queen's Secretary of State Sir Francis Walsingham (21 September). Perhaps began to revise the *Arcadia*.

1584 Worked in the Ordnance Office, with special interest in the fortification of Dover. Appointed to abortive diplomatic mission to France (July). Wrote a *Defence* of his uncle, the Earl of Leicester, in response to a libel on him (*Misc. Prose*, 123–41).

1585 Unfinished literary projects on which he may have worked in the earlier part of the year include the 'New' *Arcadia* and translations of the Psalms, of Du Plessis Mornay's *De la verité de la religion chrestienne*, and of Du Bartas's *La semaine ou Creation du monde* (1578). Appointed Joint Master of the Ordnance (July). Travelled to Plymouth with Sir Francis Drake in the hope of accompanying him to the West Indies (September); peremptorily recalled by the Queen, who appointed him governor of the cautionary town of Flushing, in the Netherlands; Leicester had just been appointed lieutenant-general of the English troops sent out to support the Dutch provinces which had revolted against Spanish rule. Daughter Elizabeth born. Arrived in Flushing (November).

1586 Leicester enraged the Queen by accepting the title of Governor General of the Netherlands (January). Sir Henry Sidney died at Worcester (May). Sidney made a successful night attack on the town of Axel (July). Lady Mary Sidney died (August). Leicester, Sidney, and others made an early morning raid on a Spanish baggage train approaching the besieged city of Zutphen in which Sidney was wounded with a musket shot in the thigh (22 September); he died of infection at Arnhem (17 October).

1587 Funeral at St Paul's (16 February), eight days after the execution of Mary, Queen of Scots. Publication of Arthur Golding's translation of Du Plessis Mornay, claimed as a completion of Sidney's.

1590 First publication of the 'New' *Arcadia*, the manuscript of which Sidney had entrusted to Greville in 1585.

1591 First publication of *Astrophil and Stella*, in a poor text, with a preface by Thomas Nashe.

1593 Publication of 'New' *Arcadia* with Books 3–5 of the 'Old' version appended.

1595 Publication of *The Defence of Poesy* (two editions, one with title *An apology for poetry*).

1598 Publication of the first 'complete Sidney', probably under the supervision of the Countess of Pembroke, including *CS*, *AS*, *DP*, and *LM* in addition to the composite *Arcadia*. Ten further editions during the seventeenth century.

1599 The Countess of Pembroke completed her translation of the Psalms, left unfinished by her brother.

1652 Fulke Greville's *Life of Sidney* (composed *c.*1610–14) published.

1907 Bertram Dobell discovered that *OA* survived in complete form in manuscript. Nine manuscripts are now known.

Note on the Text

Texts of Sidney's literary works are based when possible on previous Oxford editions. Modernization, where necessary, and some regularization of punctuation, have been carried out silently, but substantive emendations are recorded in the Notes.

Sidney's poems are referred to according to the system of reference devised by Ringler: see Abbreviations.

THE LADY OF MAY

Her most excellent Majesty walking in Wanstead Garden, as she passed down into the grove, there came suddenly among the train one apparelled like an honest man's wife of the country; where, crying out for justice, and desiring all the lords and gentlemen to speak a good word for her, she was brought to the presence of her Majesty, to whom upon her knees she offered a supplication, and used this speech:

The Suitor: Most fair lady; for as for other your titles of state, statelier persons shall give you, and thus much mine own eyes are witnesses of: take here the complaint of me, poor wretch, as deeply
10 plunged in misery, as I wish to you the highest point of happiness. One only daughter I have, in whom I placed all the hopes of my good hap, so well had she with her good parts recompensed my pain of bearing her, and care of bringing her up. But now, alas, that she is come to the time that I should reap my full comfort of her, so is she troubled with that notable matter, which we in the country call matrimony, as I cannot choose but fear the loss of her wits, at least of her honesty. Other women think they may be unhappily cumbered with one master husband; my poor daughter is oppressed with two, both loving her, both equally liked of her, both striving
20 to deserve her. But now lastly (as this jealousy, forsooth, is a vile matter) each have brought their partakers with them, and are at this present, without your presence redress it, in some bloody controversy; my poor child is among them. Now, sweet lady, help; your own way guides you to the place where they encumber her. I dare stay here no longer, for our men say here in the country, the sight of you is infectious.

And with that she went away a good pace, leaving the supplication with her Majesty, which very formally contained this:

Supplication

30 *Most gracious Sovereign:*

To one whose state is raised over all,
Whose face doth oft the bravest sort enchant,
Whose mind is such, as wisest minds appal,
Who in one self these diverse gifts can plant:

How dare I, wretch, seek there my woes to rest,
Where ears be burnt, eyes dazzled, hearts oppressed?
Your state is great, your greatness is our shield,
Your face hurts oft, but still it doth delight,
Your mind is wise, your wisdom makes you mild;
Such planted gifts enrich even beggars' sight: 40
So dare I, wretch, my bashful fear subdue,
And feed mine ears, mine eyes, mine heart in you.

Herewith, the woman suitor being gone, there was heard in the woods a confused noise, and forthwith there came out six shepherds, with as many fosters, haling and pulling to whether side they should draw the Lady of May, who seemed to incline neither to the one nor other side. Among them was Master Rombus, a schoolmaster of a village thereby, who, being fully persuaded of his own learned wisdom, came thither with his authority to part their fray; where for an answer he received many unlearned blows. But the Queen coming to the place, where she was seen of them, though 50 *they knew not her estate, yet something there was which made them startle aside and gaze upon her: till old father Lalus stepped forth (one of the substantiallest shepherds) and making a leg or two, said these few words:*

Lalus the old shepherd. May it please your benignity to give a little superfluous intelligence to that which, with the opening of my mouth, my tongue and teeth shall deliver unto you. So is it, right worshipful audience, that a certain she-creature, which we shepherds call a woman, of a minsical countenance, but by my white lamb, not three quarters so beauteous as yourself, hath disanulled the brain-pan of two of our featioust young men. And will you wot 60 how? By my mother Kit's soul, with a certain fransical malady they called 'love'; when I was a young man they called it flat folly. But here is a substantial schoolmaster can better disnounce the whole foundation of the matter, although in sooth, for all his loquence our young men were nothing duteous to his clerkship. Come on, come on, Master Rombus, be not so bashless; we say that the fairest are ever the gentlest. Tell the whole case, for you can much better vent the points of it than I.

Then came forward Master Rombus, and with many special graces made this learned oration: 70

Now the thunderthumping Jove transfund his dotes into your excellent formosity, which have with your resplendent beams thus segregated the enmity of these rural animals. I am,

Potentissima Domina, a schoolmaster; that is to say, a pedagogue; one not a little versed in the disciplinating of the juvental fry, wherein (to my laud I say it) I use such geometrical proportion, as neither wanteth mansuetude nor correction, for so it is described:

Parcere subjectis et debellare superbos.

80 Yet hath not the pulchritude of my virtues protected me from the contaminating hands of these plebeians; for coming, *solummodo*, to have parted their sanguinolent fray, they yielded me no more reverence than if I had been some *Pecorius Asinus*: I, even I, that am, who I am. *Dixi. Verbum sapiento satum est.* But what said that Trojan Aeneas, when he sojourned in the surging sulks of the sandiferous seas: *Haec olim meminisse iuvabit*. Well well, *ad propositos revertebo*; the purity of the verity is, that a certain *pulchra puella profectò*, elected and constituted by the integrated determination of all this topographical region, as the sovereign lady of this, Dame 90 Maia's month, hath been *quodammodo* hunted, as you would say, pursued, by two, a brace, a couple, a cast of young men, to whom the crafty coward Cupid had *inquam* delivered his dire doleful digging dignifying dart.

But here the May Lady interrupted his speech, saying to him:

Away, away you tedious fool, your eyes are not worthy to look to yonder princely sight, much less your foolish tongue to trouble her wise ears.

At which Master Rombus in a great chafe cried out:

O Tempori, O Moribus! In profession a child, in dignity a woman, 100 *in ceteris* a maid, should thus turpify the reputation of my doctrine with the superscription of a fool! *O Tempori, O Moribus!*

But here again the May Lady, saying to him:

Leave off, good Latin fool, and let me satisfy the long desire I have had to feed mine eyes with the only sight this age hath granted to the world.

The poor schoolmaster went his way back, and the lady kneeling down said in this manner:

Do not think, sweet and gallant lady, that I do abase myself thus much unto you because of your gay apparel; for what is so brave as the natural beauty of the flowers? nor because a certain 110 gentleman hereby seeks to do you all the honour he can in this house; that is not the matter; he is but our neighbour, and these be our own groves; nor yet because of your great estate, since no estate can be compared to be the Lady of the whole month of May, as I am. So that since both this place and this time are my servants, you may be sure I would look for reverence at your hands, if I did not see something in your face which makes me yield to you. The truth is, you excel me in that wherein I desire most to excel, and that makes me give this homage unto you, as the beautifullest lady these woods have ever received. But now, as old father Lalus directed 120 me, I will tell you my fortune, that you may be judge of my mishaps, and others' worthiness. Indeed so it is that I am a fair wench, or else I am deceived, and therefore by the consent of all our neighbours have been chosen for the absolute Lady of this merry month. With me have been (alas I am ashamed to tell it) two young men, the one a forester named Therion, the other Espilus a shepherd, very long even in love forsooth. I like them both, and love neither. Espilus is the richer, but Therion the livelier. Therion doth me many pleasures, as stealing me venison out of these forests, and many other such like pretty and prettier services; but withal he 130 grows to such rages, that sometimes he strikes me, sometimes he rails at me. This shepherd, Espilus, of a mild disposition, as his fortune hath not been to do me great service, so hath he never done me any wrong; but feeding his sheep, sitting under some sweet bush, sometimes, they say, he records my name in doleful verses. Now the question I am to ask you, fair lady, is whether the many deserts and many faults of Therion, or the very small deserts and no faults of Espilus be to be preferred. But before you give your judgement (most excellent lady) you shall hear what each of them can say for themselves in rural songs. 140

Thereupon Therion challenged Espilus to sing with him, speaking these six verses:

Therion: Come, Espilus, come now declare thy skill,
Show how thou canst deserve so brave desire,
Warm well thy wits, if thou wilt win her will,
For water cold did never promise fire:

Great sure is she, on whom our hopes do live,
Greater is she, who must the judgement give.

But Espilus, as if he had been inspired by the Muses, began forthwith to
150 *sing, whereto his fellow shepherds set in with their recorders, which they bare in their bags like pipes; and so of Therion's side did the foresters, with the cornets they wore about their necks like hunting horns in baldrics.*

Espilus: Tune up, my voice, a higher note I yield:
To high conceits the song must needs be high;
More high than stars, more firm than flinty field
Are all my thoughts, in which I live or die:
Sweet soul, to whom I vowed am a slave,
Let no wild woods so great a treasure have.

Therion: The highest note comes oft from basest mind,
160 As shallow brooks do yield the greatest sound;
Seek other thoughts thy life or death to find;
Thy stars be fall'n, ploughed is thy flinty ground:
Sweet soul, let not a wretch that serveth sheep
Among his flock so great a treasure keep.

Espilus: Two thousand sheep I have as white as milk,
Though not so white as is thy lovely face;
The pasture rich, the wool as soft as silk,
All this I give, let me possess thy grace:
But still take heed, lest thou thyself submit
170 To one that hath no wealth, and wants his wit.

Therion: Two thousand deer in wildest wood I have,
Them can I take, but you I cannot hold:
He is not poor, who can his freedom save,
Bound but to you, no wealth but you I would:
But take this beast, if beasts you fear to miss,
For of his beasts the greatest beast he is.

Espilus kneeling to the Queen:

Judge you, to whom all beauty's force is lent.

Therion: Judge you of love, to whom all love is bent.

180 *But as they waited for the judgement her Majesty should give of their deserts, the shepherds and foresters grew to a great contention whether of their fellows had sung better, and so whether the estate of shepherds or*

foresters were the more worshipful. The speakers were Dorcas, an old shepherd, and Rixus, a young forester, between whom the schoolmaster Rombus came in as moderator.

Dorcas the old shepherd: Now all the blessings of mine old grandam (silly Espilus) light upon thy shoulders for this honeycomb singing of thine. Now, by mine honesty, all the bells in the town could not have sung better. If the proud heart of the harlotry lie not down to thee now, the sheep's rot catch her, to teach her that a fair woman 190 hath not her fairness to let it grow rustish.

Rixus the foster: O Midas, why art thou not alive now to lend thine ears to this drivel? By the precious bones of a huntsman, he knows not the blaying of a calf from the song of a nightingale. But if yonder great gentlewoman be as wise as she is fair, Therion, thou shalt have the prize; and thou, old Dorcas, with young master Espilus, shall remain tame fools, as you be.

Dorcas: And with cap and knee be it spoken, is it your pleasure, neighbour Rixus, to be a wild fool?

Rixus: Rather than a sheepish dolt. 200

Dorcas: It is much refreshing to my bowels, you have made your choice; for my share, I will bestow your leavings upon one of your fellows.

Rixus: And art not thou ashamed, old fool, to liken Espilus, a shepherd, to Therion, of the noble vocation of huntsmen, in the presence of such a one as even with her eye can give the cruel punishment?

Dorcas: Hold thy peace, I will neither meddle with her nor her eyes. They say in our town, they are dangerous both. Neither will I liken Therion to my boy Espilus, since one is a thievish prowler, and the 210 other is as quiet as a lamb that new came from sucking.

Rombus the schoolmaster: *Heu, Ehem, Hei, Insipidum, Inscitium vulgorum et populorum*. Why, you brute nebulons, have you had my *corpusculum* so long among you, and cannot yet tell how to edify an argument? Attend and throw your ears to me, for I am gravidated with child, till I have indoctrinated your plumbeous cerebrosities. First you must divisionate your point, *quasi* you should cut a cheese into two particles—for thus I must uniform my speech to your obtuse conceptions; for *prius dividendum oratio antequam definiendum*,

220 *exemplum gratia*: either Therion must conquer this, Dame Maia's nymph, or Espilus must overthrow her; and that *secundum* their dignity, which must also be subdivisionated into three equal *species*, either according to the penetrancy of their singing, or the meliority of their functions, or lastly the superancy of their merits. *De* singing *satis. Nunc* are you to argumentate of the qualifying of their estate first, and then whether hath more infernally, I mean deeply, deserved.

Dorcas: O poor Dorcas, poor Dorcas, that I was not set in my young days to school, that I might have purchased the understanding of 230 Master Rombus' mysterious speeches. But yet thus much my capacity doth conceive of him, that I must give up from the bottom of my stomach what my conscience doth find in the behalf of shepherds. O sweet honey milken lambs, and is there any so flinty a heart, that can find about him to speak against them, that have the charge of so good souls as you be, among whom there is no envy, but all obedience; where it is lawful for a man to be good if he list, and hath no outward cause to withdraw him from it; where the eye may be busied in considering the works of nature, and the heart quietly rejoiced in the honest using them? If contemplation, 240 as clerks say, be the most excellent, which is so fit a life for a templer as this is, which is neither subject to violent oppression, nor servile flattery? How many courtiers, think you, I have heard under our field in bushes make their woeful complaints, some of the greatness of their mistress' estate, which dazzled their eyes and yet burned their hearts; some of the extremity of her beauty mixed with extreme cruelty; some of her too much wit, which made all their loving labours folly? O how often have I heard one name sound in many mouths, making our vales witnesses of their doleful agonies! So that with long lost labour, finding their thoughts bare 250 no other wool but despair, of young courtiers they grew old shepherds. Well, sweet lambs, I will end with you, as I began. He that can open his mouth against such innocents, let him be hated as much as a filthy fox; let the taste of him be worse than musty cheese, the sound of him more dreadful than the howling of a wolf, his sight more odible than a toad in one's porridge.

Rixius: Your life indeed hath some goodness.

Rombus the schoolmaster: O *tace, tace*, or all the fat will be ignified. First let me dilucidate the very intrinsical marrowbone of the matter. He doth use a certain rhetorical invasion into the point, as if indeed he had conference with his lambs; but the truth is, he doth equitate you in the mean time, master Rixus, for thus he saith: that sheep are good, *ergo* the shepherd is good: an *enthymeme a loco contingentibus*, as my finger and my thumb are *contingentes*. Again, he saith, who liveth well is likewise good but shepherds live well, *ergo* they are good; a syllogism in Darius King of Pesia *a coniugatis*: as you would say, a man coupled to his wife, two bodies but one soul. But do you but acquiescate to my exhortation, and you shall extinguish him. Tell him his *major* is a knave, his *minor* is a fool, and his conclusion both: *et ecce homo blancatus quasi lilium*.

Rixus: I was saying the shepherd's life had some goodness in it, because it borrowed of the country quietness something like ours. But that is not all; for ours, besides that quiet part, doth both strengthen the body, and raise up the mind with this gallant sort of activity. O sweet contentation, to see the long life of the hurtless trees; to see how in straight growing up, though never so high, they hinder not their fellows; they only enviously trouble, which are crookedly bent. What life is to be compared to ours, where the very growing things are ensamples of goodness? We have no hopes, but we may quickly go about them, and going about them, we soon obtain them; not like those that, having long followed one (in truth) most excellent chase, do not at length perceive she could never be taken; but that if she stayed at any time near her pursuers, it was never meant to tarry with them, but only to take breath to fly further from them. He therefore that doubts that our life doth not far excel all others, let him also doubt that the well deserving and painful Therion is not to be preferred before the idle Espilus, which is even as much to say, as that roes are not swifter than sheep, nor stags more goodly than goats.

Rombus: *Bene, bene, nunc de questione propositus*: that is as much to say, as well, well, now of the proposed question, that was, whether the many great services and many great faults of Therion, or the few small services and no faults of Espilus, be to be preferred, incepted or accepted the former.

The May Lady: No, no, your ordinary brains shall not deal in that matter; I have already submitted it to one whose sweet spirit hath passed through greater difficulties, neither will I that your block-heads lie in her way. Therefore, O lady, to see the accomplishment

of your desires, since all your desires be most worthy of you, vouchsafe our ears such happiness, and me that particular favour,
300 as you will judge whether of these two be more worthy of me, or whether I be worthy of them; and this I will say, that in judging me, you judge more than me in it.

This being said, it pleased her Majesty to judge that Espilus did the better deserve her; but what words, what reasons she used for it, this paper, which carrieth so base names, is not worthy to contain. Sufficeth it that upon the judgement given, the shepherds and foresters made a full consort of their cornets and recorders, and then did Espilus sing this song, tending to the greatness of his own joy, and yet to the comfort of the other side, since they were overthrown by a most worthy adversary. The song
310 *contained two short tales, and thus it was:*

Espilus: Sylvanus long in love, and long in vain,
At length obtained the point of his desire,
When being asked, now that he did obtain
His wished weal, what more he could require:
'Nothing', said he, 'for most I joy in this,
That goddess mine, my blessed being sees.'

Therion: When wanton Pan, deceived with lion's skin,
Came to the bed, where wound for kiss he got,
To woe and shame the wretch did enter in,
320 Till this he took, for comfort of his lot:
'Poor Pan', he said, 'although thou beaten be,
It is no shame, since Hercules was he.'

Espilus: Thus joyful I in chosen tunes rejoice,
That such a one is witness of my heart,
Whose clearest eyes I bliss, and sweetest voice,
That see my good, and judgeth my desert.

Therion: Thus woeful I in woe this salve do find,
My foul mishap came yet from fairest mind.

The music fully ended, the May Lady took her leave in this sort:

330 Lady your self, for other titles do rather diminish than add unto you: I and my little company must now leave you. I should do you wrong to beseech you to take our follies well, since your bounty is such as to pardon greater faults. Therefore I will wish you good night, praying to God, according to the title I possess, that as

hitherto it hath excellently done, so henceforward the flourishing of May, may long remain in you and with you.

And so they parted, leaving Master Rombus, who presented her Majesty with a chain of round agates something like beads, first beginning in a chafe:

Videte these obscure barbarons, *perfidem perfide*; you were well 340 served to be vapilated, relinquishing my dignity before I have valedixed this nymph's serenity. Well, *alias* I will be vindicated. But to you, Juno, Venus, Pallas *et profecto plus*, I have to ostend a mellifluous fruit of my fidelity. *Sic est*, so it is, that in this our city we have a certain neighbour, they call him Master Robert of Wanstead. He is counted an honest man, and one that loves us doctified men *pro vita*; and when he comes to his ædicle he distributes *oves, boves et pecora campi* largely among the *populorum*. But so stays the case, that he is foully commaculated with the papistical enormity, *O heu Aedipus Aecastor*. The *bonus vir* is a huge 350 *catholicam*, wherewith my conscience being replenished, could no longer refrain it from you, *proba dominus doctor, probo inveni*. I have found *unum par*, a pair, *papisticorum bedorum*, of Papistian beads, *cum quos*, with the which, *omnium dierum*, every day, next after his *pater noster* he *semper* saith 'and Elizabeth', as many lines as there be beads on this string. *Quamobrem*, I say, *secundum* the civil law, nine hundredth paragroper of the 7. ii. code in the great Turk Justinian's library, that he hath deponed all his juriousdiction therein, and it is forfeited *tibi dominorum domina: accipe* therefore, for he will never be so audacious to reclamat it again, being *iure* 360 *gentiorum* thus manumissed. Well, *vale, vale, felissima, et me ut facias ama*: that is, to love me much better than you were wont. And so *iterum valeamus et plauditamus*: PLAUDITAMUS ET VALEAMUS.

POEMS FROM *CERTAIN SONNETS*

Song: 'Sleep, baby mine, desire'

To the tune of *Basciami vita mia*

Sleep, baby mine, desire; nurse beauty singeth;
Thy cries, O baby, set mine head on aching:
The babe cries: 'Way, thy love doth keep me waking.'

Lully, lully, my babe; hope cradle bringeth,
Unto my children alway good rest taking:
The babe cries: 'Way, thy love doth keep me waking.'

Since, baby mine, from me thy watching springeth;
Sleep then a little, pap content is making:
The babe cries: 'Nay, for that abide I waking.'

Song: 'Who hath his fancy pleased'

To the tune of *Wilhelmus van Nassouwe &c.*

Who hath his fancy pleased
With fruits of happy sight,
Let here his eyes be raised
On nature's sweetest light:
A light which doth dissever
And yet unite the eyes,
A light which, dying never,
Is cause the looker dies.

She never dies, but lasteth
In life of lover's heart; 10
He ever dies, that wasteth
In love his chiefest part.
Thus is her life still guarded
In never dying faith;

Thus is his death rewarded,
Since she lives in his death.

Look then, and die; the pleasure
Doth answer well the pain;
Small loss of mortal treasure,
Who may immortal gain. 20
Immortal be her graces,
Immortal is her mind;
They fit for heavenly places,
This heaven in it doth bind.

But eyes those beauties see not,
Nor sense that grace descries;
Yet eyes deprived be not
From sight of her fair eyes;
Which, as of inward glory
They are the outward seal, 30
So may they live still sorry
Which die not in that weal.

But who hath fancies pleased
With fruits of happy sight
Let here his eyes be raised
On nature's sweetest light.

'Ring out your bells'

Ring out your bells, let mourning shows be spread,
For love is dead:
All love is dead, infected
With plague of deep disdain,
Worth as naught worth rejected,
And faith fair scorn doth gain.
From so ungrateful fancy,
From such a female franzy,
From them that use men thus:
Good lord, deliver us. 10

Weep, neighbours, weep: do you not hear it said
That love is dead?
His death-bed peacock's folly,
His winding-sheet is shame,
His will false-seeming holy,
His sole executor blame.
From so ungrateful fancy,
From such a female franzy,
From them that use men thus:
God lord, deliver us. 20

Let dirge be sung, and trentals rightly read,
For love is dead.
Sir wrong his tomb ordaineth,
My mistress' marble heart,
Which epitaph containeth:
'Her eyes were once his dart.'
From so ungrateful fancy,
From such a female franzy,
From them that use men thus:
Good lord, deliver us. 30

Alas, I lie: rage hath this error bred;
Love is not dead.
Love is not dead, but sleepeth
In her unmatched mind,
Where she his counsel keepeth
Till due desert she find.
Therefore from so vile fancy,
To call such wit a franzy
Who love can temper thus:
Good lord, deliver us. 40

Sonnet: 'Thou blind man's mark'

Thou blind man's mark, thou fool's self-chosen snare,
Fond fancy's scum, and dregs of scattered thought,
Band of all evils, cradle of causeless care,
Thou web of will, whose end is never wrought;

Desire, desire, I have too dearly bought,
With price of mangled mind, thy worthless ware;
Too long, too long, asleep thou hast me brought,
Who should my mind to higher things prepare.
  But yet in vain thou hast my ruin sought
  In vain thou madest me to vain things aspire, 10

  In vain thou kindlest all thy smoky fire;
  For virtue hath this better lesson taught,
  Within myself to seek my only hire,
  Desiring naught but how to kill desire.

Sonnet: 'Leave me, O love'

Leave me, O love which reachest but to dust,
And thou, my mind, aspire to higher things;
Grow rich in that which never taketh rust;
What ever fades, but fading pleasure brings.
Draw in thy beams, and humble all thy might
To that sweet yoke where lasting freedoms be,
Which breaks the clouds, and opens forth the light
That doth both shine, and give us sight to see.
O take fast hold, let that light be thy guide
In this small course which birth draws out to death, 10
And think how ill becometh him to slide,
Who seeketh heaven, and comes of heavenly breath:
  Then farewell, world; thy uttermost I see;
  Eternal love, maintain thy life in me.

Splendidis longum valedico nugis

FROM 1593 *ARCADIA*

'The lad Philisides'

The lad Philisides
Lay by a river's side,
In flowery field a gladder eye to please;
His pipe was at his foot,
His lambs were him beside;
A widow turtle near on bared root
Sate wailing without boot;
Each thing, both sweet and sad,
Did draw his boiling brain
To think, and think with pain, 10
Of Mira's beams, eclipsed by absence bad.
And thus, with eyes made dim
With tears, he said, or sorrow said for him:

'O earth, once answer give:
So may thy stately grace
By north or south still rich adorned live;
So Mira long may be
On thy then blessed face,
Whose foot doth set a heaven on cursed thee;
I ask—now answer me— 20
If th'author of thy bliss,
Phoebus, that shepherd high,
Do turn from thee his eye,
Doth not thyself, when he long absent is,
Like rogue all ragged go,
And pine away with daily wasting woe?

'Tell me, you wanton brook:
So may your sliding race
Shun loathed-loving banks with cunning crook;
So in you ever new 30
Mira may look her face,
And make you fair with shadow of her hue,

So when to pay your due
To mother sea you come,
She chide you not for stay,
Nor beat you for your play:
Tell me, if your diverted streams become
Absented quite from you,
Are you not dried? Can you yourself renew?

'Tell me, you flowers fair, 40
Cowslip and columbine:
So may your make, this wholesome spring-time air,
With you embraced lie,
And lately thence untwine,
But with dew-drops engender children high;
So may you never die,
But pulled by Mira's hand
Dress bosom hers, or head,
Or scatter on her bed:
Tell me, if husband spring-time leave your land, 50
When he from you is sent,
Wither not you, languished with discontent?

'Tell me, my seely pipe:
So may thee still betide
A cleanly cloth thy moistness for to wipe;
So may the cherries red
Of Mira's lips divide
Their sugared selves to kiss thy happy head;
So may her ears be led,
Her ears, where music lives, 60
To hear, and not despise,
Thy liribliring cries:
Tell, if that breath which thee thy sounding gives
Be absent far from thee,
Absent alone canst thou then piping be?

'Tell me, my lamb of gold:
So may'st thou long abide
The day well fed, the night in faithful fold;
So grow thy wool of note
In time, that, richly dyed, 70

It may be part of Mira's petticoat;
Tell me, if wolves the throat
Have caught of thy dear dam,
Or she from thee be stayed,
Or thou from her be strayed,
Canst thou, poor lamb, become another's lamb?
Or rather, till thou die,
Still for thy dam with bea-waymenting cry?

'Tell me, O turtle true:
So may no fortune breed 80
To make thee, nor thy better-loved, rue;
So may thy blessings swarm
That Mira may thee feed
With hand and mouth; with lap and breast keep warm:
Tell me, if greedy arm
Do fondly take away
With traitor lime the one,
The other left alone;
Tell me, poor wretch, parted from wretched prey,
Disdain not you the green, 90
Wailing till death; shun you not to be seen?

'Earth, brook, flowers, pipe, lamb, dove,
Say all, and I with them:
'Absence is death, or worse, to them that love'.
So I, unlucky lad,
Whom hills from her do hem,
What fits me now but tears, and sighings sad?
O fortune too too bad:
I rather would my sheep
Th'had'st killed with a stroke, 100
Burnt cabin, lost my cloak,
Than want one hour those eyes which my joys keep.
O, what doth wailing win?
Speech without end were better not begin.

'My song, climb thou the wind
Which Holland sweet now gently sendeth in,
That on his wings the level thou may'st find
To hit, but kissing hit,

Her ears, the weights of wit.
If thou know not for whom thy master dies. 110
These marks shall make thee wise:
She is the herdess fair that shines in dark,
And gives her kids no food but willow's bark.'

This said, at length he ended
His oft sigh-broken ditty,
Then rase; but rase on legs with faintness bended,
With skin in sorrow dyed,
With face the plot of pity,
With thoughts, which thoughts their own tormentors tried,
He rase, and straight espied 120
His ram, who to recover
The ewe another loved
With him proud battle proved:
He envied such a death in sight of lover,
And always westward eyeing,
More envied Phoebus for his western flying.

POEMS FROM THE 'OLD' *ARCADIA*

'What tongue can her perfections tell'

What tongue can her perfections tell
In whose each part all pens may dwell?
Her hair fine threads of finest gold
In curled knots man's thoughts to hold;
But that her forehead says, 'in me
A whiter beauty you may see.'
Whiter indeed; more white than snow
Which on cold winter's face doth grow.
  That doth present those even brows,
Whose equal lines their angles bows, 10
Like to the moon when after change
Her horned head abroad doth range;
And arches be to heav'nly lids,
Whose wink each bold attempt forbids.
  For the black stars those spheres contain,
Their matchless praise, e'en praise doth stain.
No lamp whose light by art is got,
No sun which shines, and seeth not,
Can liken them without all peer,
Save one as much as other clear; 20
Which only thus unhappy be
Because themselves they cannot see.
  Her cheeks with kindly claret spread,
Aurora-like new out of bed,
Or like the fresh queen-apple's side,
Blushing at sight of Phoebus' pride.
Her nose, her chin, pure ivory wears,
No purer than the pretty ears,
So that therein appears some blood,
Like wine and milk that mingled stood. 30
In whose incirclets if you gaze
Your eyes may tread a lover's maze,
But with such turns the voice to stray,
No talk untaught can find the way.

The tip no jewel needs to wear;
The tip is jewel of the ear.
  But who those ruddy lips can miss,
Which blessed still themselves do kiss?
Rubies, cherries, and roses new,
In worth, in taste, in perfect hue, 40
Which never part but that they show
Of precious pearl the double row,
The second sweetly-fenced ward
Her heav'nly-dewed tongue to guard,
Whence never word in vain did flow.
  Fair under these doth stately grow
The handle of this pleasant work,
The neck, in which strange graces lurk.
Such be, I think, the sumptuous towers
Which skill doth make in princes' bowers. 50
  So good a say invites the eye
A little downward to espy
The lovely clusters of her breasts,
Of Venus' babe the wanton nests,
Like pommels round of marble clear,
Where azured veins well mixed appear,
With dearest tops of porphyry.
  Betwixt these two a way doth lie,
A way more worthy beauty's fame
Than that which bears the milken name. 60
This leads unto the joyous field
Which only still doth lilies yield;
But lilies such whose native smell
The Indian odours doth excel.
Waist it is called, for it doth waste
Men's lives until it be embraced.
  There may one see, and yet not see,
Her ribs in white well armed be,
More white than Neptune's foamy face
When struggling rocks he would embrace. 70
  In these delights the wand'ring thought
Might of each side astray be brought,
But that her navel doth unite
In curious circle busy sight,
A dainty seal of virgin wax

Where nothing but impression lacks.
   Her belly there glad sight doth fill,
Justly entitled Cupid's hill;
A hill most fit for such a master,
A spotless mine of alabaster, 80
Like alabaster fair and sleek,
But soft and supple, satin-like,
In that sweet seat the boy doth sport.
Loath, I must leave his chief resort;
For such an use the world hath gotten,
The best things still must be forgotten.
   Yet never shall my song omit
Those thighs (for Ovid's song more fit)
Which, flanked with two sugared flanks,
Lift up their stately swelling banks 90
That Albion cliffs in whitness pass,
With haunches smooth as looking glass.
   But bow all knees, now of her knees
My tongue doth tell what fancy sees:
The knots of joy, the gems of love,
Whose motion makes all graces move;
Whose bought incaved doth yield such sight,
Like cunning painter shadowing white.
The gart'ring place with childlike sign
Shows easy print in metal fine. 100
   But there again the flesh doth rise
In her brave calves like crystal skies,
Whose Atlas is a smallest small,
More white than whitest bone of whale.
   There oft steals out that round clean foot,
This noble cedar's precious root;
In show and scent pale violets,
Whose step on earth all beauty sets.
   But back unto her back, my muse,
Where Leda's swan his feathers mews, 110
Along whose ridge such bones are met,
Like comfits round in marchpane set.
   Her shoulders be like two white doves,
Perching within square royal rooves,
Which leaded are with silver skin,
Passing the hate-spot ermelin.

And thence those arms derived are;
The phoenix' wings be not so rare
For faultless length and stainless hue.
Ah, woe is me, my woes renew! 120
Now course doth lead me to her hand,
Of my first love the fatal band,
Where whiteness doth for ever sit;
Nature herself enamelled it.
For there with strange compact doth lie
Warm snow, moist pearl, soft ivory.
There fall those sapphire-coloured brooks,
Which conduit-like, with curious crooks,
Sweet islands make in that sweet land.
As for the fingers of the hand 130
The bloody shafts of Cupid's war,
With amethysts they headed are.
Thus hath each part his beauty's part;
But how the Graces do impart
To all her limbs a special grace,
Becoming every time and place,
Which doth e'en beauty beautify,
And most bewitch the wretched eye!
How all this is but a fair inn
Of fairer guest which dwells within, 140
Of whose high praise, and praiseful bliss,
Goodness the pen, heav'n paper is;
The ink immortal fame doth lend.
As I began, so must I end:
No tongue can her perfections tell,
In whose each part all pens may dwell.

Epithalamium: 'Let mother earth'

Let mother earth now deck herself in flowers,
To see her offspring seek a good increase,
Where justest love doth vanquish Cupid's powers
And war of thoughts is swallowed up in peace
Which never may decrease,
But like the turtles fair

Live one in two, a well united pair,
Which, that no chance may stain,
O Hymen long their coupled joys maintain.

O heav'n awake, show forth thy stately face; 10
Let not these slumb'ring clouds thy beauties hide,
But with thy cheerful presence help to grace
The honest bridegroom and the bashful bride,
Whose loves may ever bide,
Like to the elm and vine,
With mutual embracements them to twine;
In which delightful pain,
O Hymen long their coupled joys maintain.

Ye muses all which chase affects allow,
And have to Lalus showed your secret skill, 20
To this chaste love your sacred favours bow,
And so to him and her your gifts distil,
That they all vice may kill;
And like to lilies pure
Do please all eyes, and spotless do endure;
Where, that all bliss may reign,
O Hymen long their coupled joys maintain.

Ye nymphs which in the waters empire have,
Since Lalus' music oft doth yield you praise,
Grant to the thing which we for Lalus crave: 30
Let one time (but long first) close up their days,
One grave their bodies seize,
And like two rivers sweet
When they, though diverse, do together meet,
One stream both streams contain;
O Hymen long their coupled joys maintain.

Pan, father Pan, the god of silly sheep,
Whose care is cause that they in number grow,
Have much more care of them that them do keep,
Since from these good the others' good doth flow, 40
And make their issue show
In number like the herd
Of younglings which thyself with love hast reared,

Or like the drops of rain;
O Hymen long their coupled joys maintain.

Virtue, if not a god, yet God's chief part,
Be thou the knot of this their open vow:
That still he be her head, she be his heart,
He lean to her, she unto him do bow;
Each other still allow, 50
Like oak and mistletoe,
Her strength from him, his praise from her do grow.
In which most lovely train,
O Hymen long their coupled joys maintain.

But thou foul Cupid, sire to lawless lust,
Be thou far hence with thy empoisoned dart
Which, though of glitt'ring gold, shall here take rust
Where simple love, which chasteness doth impart,
Avoids thy hurtful art,
Not needing charming skill 60
Such minds with sweet affections for to fill,
Which being pure and plain,
O Hymen long their coupled joys maintain.

All churlish words, shrewd answers, crabbed looks,
All privateness, self-seeking, inward spite,
All waywardness which nothing kindly brooks,
All strife for toys, and claiming master's right,
Be hence ay put to flight;
All stirring husband's hate
Gainst neighbours good for womanish debate 70
Be fled as things most vain,
O Hymen long their coupled joys maintain.

All peacock pride, and fruits of peacock's pride,
Longing to be with loss of substance gay
With recklessness what may thy house betide,
So that you may on higher slippers stay,
For ever hence away.
Yet let not sluttery,
The sink of filth, be counted housewifery;
But keeping wholesome mean, 80
O Hymen long their coupled joys maintain.

But above all, away vile jealousy,
The ill of ills, just cause to be unjust,
(How can he love, suspecting treachery?
How can she love where love cannot win trust?)
  Go snake, hide thee in dust,
  Ne dare once show thy face
  Where open hearts do hold so constant place;
  That they thy sting restrain,
  O Hymen long their coupled joys maintain. 90

The earth is decked with flow'rs, the heav'ns displayed,
Muses grant gifts, nymphs long and joined life,
Pan store of babes, virtue their thoughts well stayed,
Cupid's lust gone, and gone is bitter strife,
  Happy man, happy wife.
  No pride shall them oppress,
  Nor yet shall yield to loathsome sluttishness,
  And jealousy is slain;
  For Hymen will their coupled joys maintain.

Philisides' fable: 'As I my little flock'

As I my little flock on Ister bank
(A little flock, but well my pipe they couthe)
Did piping lead, the sun already sank
Beyond our world, and ere I gat my booth
Each thing with mantle black the night doth soothe,
  Saving the glow-worm, which would courteous be
  Of that small light oft watching shepherds see.

The welkin had full niggardly enclosed
In coffer of dim clouds his silver groats,
Ycleped stars; each thing to rest disposed: 10
The caves were full, the mountains void of goats;
The birds' eyes closed, closed their chirping notes.
  As for the nightingale, wood-music's king,
  It August was, he deigned not then to sing.

Amid my sheep, though I saw naught to fear,
Yet (for I nothing saw) I feared sore;
Then found I which thing is a charge to bear,
For for my sheep I dreaded mickle more
Than ever for myself since I was bore:
  I sat me down, for see to go ne could, 20
  And sang unto my sheep lest stray they should.

The song I sang old Languet had me taught,
Languet, the shepherd best swift Ister knew,
For clerkly rede, and hating what is naught,
For faithful heart, clean hands, and mouth as true.
With his sweet skill my skill-less youth he drew
  To have a feeling taste of him that sits
  Beyond the heav'n, far more beyond our wits.

He said the music best thilk powers pleased
Was jump concord between our wit and will, 30
Where highest notes to godliness are raised,
And lowest sink not down to jot of ill.
With old true tales he wont mine ears to fill:
  How shepherds did of yore, how now they thrive,
  Spoiling their flock, or while twixt them they strive.

He liked me, but pitied lustful youth.
His good strong staff my slipp'ry years upbore.
He still hoped well, because I loved truth;
Till forced to part, with heart and eyes e'en sore,
To worthy Coredens he gave me o'er. 40
  But thus in oak's true shade recounted he
  Which now in night's deep shade sheep heard of me.

Such manner time there was (what time I not)
When all this earth, this dam or mould of ours,
Was only woned with such a beasts begot;
Unknown as then were they that builden towers.
The cattle, wild or tame, in nature's bowers
  Might freely roam or rest, as seemed them;
  Man was not man their dwellings in to hem.

The beasts had sure some beastly policy; 50
For nothing can endure where order nis.
For once the lion by the lamb did lie;
The fearful hind the leopard did kiss;
Hurtless was tiger's paw and serpent's hiss.
  This think I well: the beasts with courage clad
  Like senators a harmless empire had.

At which, whether the others did repine
(For envy harb'reth most in feeblest hearts),
Or that they all to changing did incline
(As e'en in beasts their dams leave changing parts), 60
The multitude to Jove a suit imparts,
  With neighing, blaying, braying, and barking,
  Roaring, and howling, for to have a king.

A king in language theirs they said they would
(For then their language was a perfect speech).
The birds likewise with chirps and pewing could,
Cackling and chatt'ring, that of Jove beseech.
Only the owl still warned them not to seech
  So hastily that which they would repent;
  But saw they would, and he to deserts went. 70

Jove wisely said (for wisdom wisely says):
'O beasts, take heed what you of me desire.
Rulers will think all things made them to please,
And soon forget the swink due to their hire.
But since you will, part of my heav'nly fire
  I will you lend; the rest yourselves must give,
  That it both seen and felt may with you live.'

Full glad they were, and took the naked sprite,
Which straight the earth yclothed in his clay.
The lion, heart; the ounce gave active might; 80
The horse, good shape; the sparrow, lust to play;
Nightingale, voice, enticing songs to say.
  Elephant gave a perfect memory;
  And parrot, ready tongue, that to apply.

The fox gave craft; the dog gave flattery;
Ass, patience; the mole, a working thought;

Eagle, high look; wolf, secret cruelty;
Monkey, sweet breath; the cow, her fair eyes brought;
The ermine, whitest skin spotted with naught;
  The sheep, mild-seeming face; climbing, the bear; 90
  The stag did give the harm-eschewing fear.

The hare her sleights, the cat his melancholy;
Ant, industry; and cony, skill to build;
Cranes, order; storks, to be appearing holy;
Chameleon, ease to change; duck, ease to yield;
Crocodile, tears which might be falsely spilled.
  Ape great thing gave, though he did mowing stand:
  The instrument of instruments, the hand.

Each other beast likewise his present brings;
And (but they drad their prince they oft should want) 100
They all consented were to give him wings.
And ay more awe towards him for to plant,
To their own work this privilege they grant:
  That from thenceforth to all eternity
  No beast should freely speak, but only he.

Thus man was made; thus man their lord became;
Who at the first, wanting or hiding pride,
He did to beasts' best use his cunning frame,
With water drink, herbs meat, and naked hide,
And fellow-like let his dominion slide, 110
  Not in his sayings saying 'I', but 'we';
  As if he meant his lordship common be.

But when his seat so rooted he had found
That they now skilled not how from him to wend,
Then gan in guiltless earth full many a wound,
Iron to seek, which gainst itself should bend
To tear the bowels that good corn should send.
  But yet the common dam none did bemoan,
  Because (though hurt) they never heard her groan.

Then gan he factions in the beasts to breed; 120
Where helping weaker sort, the nobler beasts
(As tigers, leopards, bears, and lions' seed)
Disdained with this, in deserts sought their rests;

Where famine ravin taught their hungry chests,
  That craftily he forced them to do ill;
  Which being done, he afterwards would kill

For murder done, which never erst was seen,
By those great beasts. As for the weakers' good,
He chose themselves his guarders for to been
Gainst those of might of whom in fear they stood, 130
As horse and dog; not great, but gentle blood.
  Blithe were the commons, cattle of the field,
  Tho when they saw their foen of greatness killed.

But they, or spent, or made of slender might,
Then quickly did the meaner cattle find,
The great beams gone, the house on shoulders light;
For by and by the horse fair bits did bind;
The dog was in a collar taught his kind.
  As for the gentle birds, like case might rue
  When falcon they, and goshawk, saw in mew. 140

Worst fell to smallest bids, and meanest herd,
Who now his own, full like his own he used.
Yet first but wool, or feathers, off he teared;
And when they were well used to be abused,
For hungry throat their flesh with teeth he bruised;
  At length for glutton taste he did them kill;
  At last for sport their silly lives did spill.

But yet, O man, rage not beyond thy need;
Deem it no gloire to swell in tyranny.
Thou art of blood; joy not to make things bleed. 150
Thou fearest death; think they are loath to die.
A plaint of guiltless hurt doth pierce the sky.
  And you, poor beasts, in patience bide your hell,
  Or know your strengths, and then you shall do well.

Thus did I sing and pipe eight sullen hours
To sheep whom love, not knowledge, made to hear;
Now fancy's fits, now fortune's baleful stours.
But then I homeward called my lambkins dear;
For to my dimmed eyes began t'appear

The night grown old, her black head waxen grey, 160
Sure shepherd's sign that morn would soon fetch day.

Double sestina: 'Ye goat-herd gods'; followed by dizain and crown: 'I joy in grief'

Strephon Klaius

Strephon. Ye goat-herd gods, that love the grassy mountains,
Ye nymphs, which haunt the springs in pleasant valleys,
Ye satyrs, joyed with free and quiet forests,
Vouchsafe your silent ears to plaining music
Which to my woes gives still an early morning,
And draws the dolour on till weary evening.

Klaius. O Mercury, foregoer to the evening,
O heav'nly huntress of the savage mountains,
O lovely star, entitled of the morning,
While that my voice doth fill these woeful valleys, 10
Vouchsafe your silent ears to plaining music,
Which oft hath Echo tired in secret forests.

Strephon. I that was once free burgess of the forests,
Where shade from sun, and sport I sought in evening,
I that was once esteemed for pleasant music,
Am banished now among the monstrous mountains
Of huge despair, and foul affliction's valleys,
Am grown a screech-owl to myself each morning.

Klaius. I that was once delighted every morning,
Hunting the wild inhabiters of forests, 20
I that was once the music of these valleys,
So darkened am that all my day is evening,
Heart-broken so, that molehills seem high mountains,
And fill the vales with cries instead of music.

Strephon: Long since, alas, my deadly swannish music
Hath made itself a crier of the morning,
And hath with wailing strength climbed highest mountains;

Long since my thoughts more desert be than forests;
Long since I see my joys come to their evening,
And state thrown down to over-trodden valleys. 30

Klaius. Long since the happy dwellers of these valleys
Have prayed me leave my strange exclaiming music,
Which troubles their day's work, and joys of evening;
Long since I hate the night, more hate the morning;
Long since my thoughts chase me like beasts in forests,
And make me wish myself laid under mountains.

Strephon. Meseems I see the high and stately mountains
Transform themselves to low dejected valleys;
Meseems I hear in these ill-changed forests
The nightingales do learn of owls their music; 40
Meseems I feel the comfort of the morning
Turned to the mortal serene of an evening.

Klaius. Meseems I see a filthy cloudy evening
As soon as sun begins to climb the mountains;
Meseems I feel a noisome scent the morning
When I do smell the flowers of these valleys;
Meseems I hear (when I do hear sweet music)
The dreadful cries of murdered men in forests.

Strephon. I wish to fire the trees of all these forests;
I give the sun a last farewell each evening; 50
I curse the fiddling finders-out of music;
With envy I do hate the lofty mountains,
And with despite despise the humble valleys;
I do detest night, evening, day, and morning.

Klaius. Curse to myself my prayer is, the morning;
My fire is more than can be made with forests;
My state more base than are the basest valleys;
I wish no evenings more to see, each evening;
Shamed, I hate myself in sight of mountains,
And stop mine ears lest I grow mad with music. 60

Strephon. For she, whose parts maintained a perfect music,
Whose beauties shined more than the blushing morning,

Who much did pass in state the stately mountains,
In straightness passed the cedars of the forests,
Hath cast me, wretch, into eternal evening,
By taking her two suns from these dark valleys.

Klaius. For she, with whom compared the Alps are valleys,
She, whose least word brings from the spheres their music,
At whose approach the sun rase in the evening,
Who, where she went, bare in her forehead morning, 70
Is gone, is gone from these our spoiled forests,
Turning to deserts our best pastured mountains.

Strephon. These mountains witness shall, so shall these valleys,

Klaius. These forests eke, made wretched by our music,
Our morning hymn this is, and song at evening.

But, as though all this had been but the taking of a taste to their wailings, Strephon again began this dizain, which was answered unto him in that kind of verse which is called the crown:

Strephon: I joy in grief, and do detest all joys;
Despise delight, am tired with thought of ease.
I turn my mind to all forms of annoys,
And with the change of them my fancy please.
I study that which most may me displease,
And in despite of that displeasure's might
Embrace that most that most my soul destroys;
Blinded with beams, fell darkness is my sight;
Dwell in my ruins, feed with sucking smart,
I think from me, not from my woes, to part. 10

Klaius. I think from me, not from my woes, to part,
And loathe this time called life, nay think that life
Nature to me for torment did impart;
Think my hard haps have blunted death's sharp knife,
Not sparing me in whom his works be rife;
And thinking this, think nature, life, and death
Place sorrow's triumph on my conquered heart.
Whereto I yield, and seek no other breath
But from the scent of some infectious grave;
Nor of my fortune aught but mischief crave. 20

Strephon. Nor of my fortune aught but mischief crave,
And seek to nourish that which now contains
All what I am. If I myself will save,
Then must I save what in me chiefly reigns,
Which is the hateful web of sorrow's pains.
Sorrow then cherish me, for I am sorrow;
No being now but sorrow I can have;
Then deck me as thine own; thy help I borrow,
Since thou my riches art, and that thou hast
Enough to make a fertile mind lie waste. 30

Klaius. Enough to make a fertile mind lie waste
Is that huge storm which pours itself on me.
Hailstones of tears, of sighs a monstrous blast,
Thunders of cries; lightnings my wild looks be,
The darkened heav'n my soul which naught can see;
The flying sprites which trees by roots up tear
Be those despairs which have my hopes quite waste.
The difference is: all folks those storms forbear,
But I cannot; who then myself should fly,
So close unto myself my wracks do lie. 40

Strephon. So close unto myself my wracks do lie;
Both cause, effect, beginning, and the end
Are all in me: what help then can I try?
My ship, myself, whose course to love doth bend,
Sore beaten doth her mast of comfort spend;
Her cable, reason, breaks from anchor, hope;
Fancy, her tackling, torn away doth fly;
Ruin, the wind, hath blown her from her scope;
Bruised with waves of care, but broken is
On rock, despair, the burial of my bliss. 50

Klaius. On rock, despair, the burial of my bliss,
I long do plough with plough of deep desire;
The seed fast-meaning is, no truth to miss;
I harrow it with thoughts, which all conspire
Favour to make my chief and only hire.
But, woe is me, the year is gone about,
And now I fain would reap, I reap but this,
Hate fully grown, absence new sprongen out.

So that I see, although my sight impair,
Vain is their pain who labour in despair. 60

Strephon. Vain is their pain who labour in despair.
For so did I, when with my angle, will.
I sought to catch the fish torpedo fair.
E'en then despair did hope already kill;
Yet fancy would perforce employ his skill,
And this hath got: the catcher now is caught,
Lamed with the angle which itself did bear,
And unto death, quite drowned in dolours, brought
To death, as then disguised in her fair face.
Thus, thus alas, I had my loss in chase. 70

Klaius. Thus, thus alas, I had my loss in chase
When first that crowned basilisk I knew,
Whose footsteps I with kisses oft did trace,
Till by such hap as I must ever rue
Mine eyes did light upon her shining hue,
And hers on me, astonished with that sight.
Since then my heart did lose his wonted place,
Infected so with her sweet poison's might
That, leaving me for dead, to her it went.
But ah, her flight hath my dead relics spent. 80

Strephon. But ah, her flight hath my dead relics spent,
Her flight from me, from me, though dead to me,
Yet living still in her, while her beams lent
Such vital spark that her mine eyes might see.
But now those living lights absented be,
Full dead before, I now to dust should fall,
But that eternal pains my soul have hent,
And keep it still within this body thrall;
That thus I must, while in this death I dwell,
In earthly fetters feel a lasting hell. 90

Klaius. In earthly fetters feel a lasting hell
Alas I do; from which to find release,
I would the earth, I would the heavens sell.
But vain it is to think those pains should cease,
Where life is death, and death cannot breed peace.

O fair, O only fair, from thee, alas,
These foul, most foul, disasters to me fell;
Since thou from me (O me) O sun didst pass.
Therefore esteeming all good blessings toys,
I joy in grief, and do detest all joys. 100

Strephon. I joy in grief, and do detest all joys.
But now an end, O Klaius, now an end,
For e'en the herbs our hateful music stroys,
And from our burning breath the trees do bend.

Pastoral elegy: 'Since that to death'

Since that to death is gone the shepherd high
Who most the silly shepherd's pipe did prize,
Your doleful tunes sweet muses now apply.
And you, O trees (if any life there lies
In trees) now through your porous barks receive
The strange resound of these my causeful cries;
And let my breath upon your branches cleave,
My breath distinguished into words of woe,
That so I may signs of my sorrow leave.
But if among yourselves some one tree grow 10
That aptest is to figure misery,
Let it embassage bear your griefs to show.
The weeping myrrh I think will not deny
Her help to this, this justest cause of plaint.
Your doleful tunes sweet muses now apply.

And thou, poor earth, whom fortune doth attaint
In nature's name to suffer such a harm
As for to lose thy gem, our earthly saint,
Upon thy face let coaly ravens swarm;
Let all the sea thy tears accounted be; 20
Thy bowels with all killing metals arm.
Let gold now rust, let diamonds waste in thee;
Let pearls be wan with woe their dam doth bear;
Thyself henceforth the light do never see.

And you, O flow'rs, which sometimes princes were,
  Till these strange alt'rings you did hap to try,
  Of prince's loss yourselves for tokens rear.
Lily in mourning black thy whiteness dye.
  O hyacinth let ai be on thee still.
  Your doleful tunes sweet muses now apply. 30

O echo, all these woods with roaring fill,
  And do not only mark the accents last
  But all, for all reach not my wailful will;
One echo to another echo cast
  Sound of my griefs, and let it never end
  Till that it hath all woods and waters passed.
Nay, to the heav'ns your just complainings send,
  And stay the stars' inconstant constant race
  Till that they do unto our dolours bend;
And ask the reason of that special grace 40
  That they, which have no lives, should live so long,
  And virtuous souls so soon should lose their place?
Ask, if in great men good men so do throng
  That he for want of elbow-room must die?
  Or if that they be scant, if this be wrong?
Did wisdom this our wretched time espy
  In one true chest to rob all virtue's treasure?
  Your doleful tunes sweet muses now apply.

And if that any counsel you to measure
  Your doleful tunes, to them still plaining say: 50
  To well felt grief, plaint is the only pleasure.
O light of sun, which is entitled day,
  O well thou dost that thou no longer bidest;
  For mourning night her black weeds may display.
O Phoebus with good cause thy face thou hidest
  Rather than have thy all-beholding eye
  Fouled with this sight while thou thy chariot guidest.
And well (methinks) becomes this vaulty sky
  A stately tomb to cover him deceased.
  Your doleful tunes sweet muses now apply. 60

O Philomela with thy breast oppressed
  By shame and grief, help, help me to lament

Such cursed harms as cannot be redressed.
Or if thy mourning notes be fully spent,
Then give a quiet ear unto my plaining;
For I to teach the world complaint am bent.
Ye dimmy clouds, which well employ your staining
This cheerful air with your obscured cheer,
Witness your woeful tears with daily raining.
And if, O sun, thou ever did'st appear 70
In shape which by man's eye might be perceived,
Virtue is dead, now set thy triumph here.
Now set thy triumph in this world, bereaved
Of what was good, where now no good doth lie;
And by thy pomp our loss will be conceived.
O notes of mine, yourselves together tie;
With too much grief methinks you are dissolved.
Your doleful tunes sweet muses now apply.

Time ever old and young is still revolved
Within itself, and never taketh end; 80
But mankind is for ay to naught resolved.
The filthy snake her aged coat can mend,
And getting youth again, in youth doth flourish;
But unto man, age ever death doth send.
The very trees with grafting we can cherish,
So that we can long time produce their time;
But man which helpeth them, helpless must perish.
Thus, thus, the minds which over all do climb,
When they by years' experience get best graces,
Must finish then by death's detested crime. 90
We last short while, and build long-lasting places.
Ah, let us all against foul nature cry;
We nature's works do help, she us defaces.
For how can nature unto this reply:
That she her child, I say, her best child killeth?
Your doleful tunes sweet muses now apply.

Alas, methinks my weakened voice but spilleth
The vehement course of this just lamentation;
Methinks my sound no place with sorrow filleth.
I know not I, but once in detestation 100

I have myself, and all what life containeth,
Since death on virtue's fort hath made invasion.
One word of woe another after traineth;
Ne do I care how rude be my invention,
So it be seen what sorrow in me reigneth.
O elements, by whose (they say) contention
Our bodies be in living pow'r maintained,
Was this man's death the fruit of your dissension?
O physic's power, which (some say) hath refrained
Approach of death, alas thou helpest meagrely 110
When once one is for Atropos distrained.
Great be physicians' brags, but aid is beggarly;
When rooted moisture fails, or groweth dry,
They leave off all, and say death comes too eagerly.
They are but words therefore which men do buy
Of any since god Aesculapius ceased.
Your doleful tunes sweet muses now apply.

Justice, justice is now, alas, oppressed;
Bountifulness hath made his last conclusion;
Goodness for best attire in dust is dressed. 120
Shepherds bewail your uttermost confusion;
And see by this picture to you presented,
Death is our home, life is but a delusion.
For see, alas, who is from you absented.
Absented? nay, I say for ever banished
From such as were to die for him contented.
Out of our sight in turn of hand is vanished
Shepherd of shepherds, whose well settled order
Private with wealth, public with quiet, garnished.
While he did live, far, far was all disorder; 130
Example more prevailing than direction,
Far was home-strife, and far was foe from border.
His life a law, his look a full correction;
As in his health we healthful were preserved
So in his sickness grew our sure infection;
His death our death. But ah, my muse hath swerved
From such deep plaint as should such woes descry,
Which he of us for ever hath deserved.
The style of heavy heart can never fly
So high as should make such a pain notorious. 140

Cease muse, therefore; thy dart, O death, apply;
And farewell prince, whom goodness hath made glorious.

'Farewell, O sun'

Farewell O sun, Arcadia's clearest light;
Farewell O pearl, the poor man's plenteous treasure;
Farewell O golden staff, the weak man's might;
Farewell O joy, the woeful's only pleasure.
Wisdom farewell, the skill-less man's direction;
Farewell with thee, farewell all our affection.

For what place now is left for our affection,
Now that of purest lamp is queint the light
Which to our darkened minds was best direction;

Now that the mine is lost of all our treasure, 10
Now death hath swallowed up our worldly pleasure,
We orphans left, void of all public might?

Orphans indeed, deprived of father's might;
For he our father was in all affection,
In our well doing placing all his pleasure,
Still studying how to us to be a light.
As well he was in peace a safest treasure;
In war his wit and word was our direction.

Whence, whence alas, shall we seek our direction
When that we fear our hateful neighbours' might, 20
Who long have gaped to get Arcadians' treasure?
Shall we now find a guide of such affection,
Who for our sakes will think all travail light,
And make his pain to keep us safe his pleasure?

No, no, for ever gone is all our pleasure;
For ever wand'ring from all good direction;
For ever blinded of our clearest light;
For ever lamed of our surest might;
For ever banished from well placed affection;
For ever robbed of our royal treasure. 30

Let tears for him therefore be all our treasure,
And in our wailful naming him our pleasure.
Let hating of ourselves be our affection.
And unto death bend still our thoughts' direction.
Let us against ourselves employ our might,
And putting out our eyes seek we our light.

Farewell our light, farewell our spoiled treasure;
Farewell our might, farewell our daunted pleasure;
Farewell direction, farewell all affection.

ASTROPHIL AND STELLA

1

Loving in truth, and fain in verse my love to show,
That she (dear she) might take some pleasure of my pain;
Pleasure might cause her read, reading might make her know;
Knowledge might pity win, and pity grace obtain;
  I sought fit words to paint the blackest face of woe,
Studying inventions fine, her wits to entertain;
Oft turning others' leaves, to see if thence would flow
Some fresh and fruitful showers upon my sunburnt brain.
  But words came halting forth, wanting invention's stay;
Invention, nature's child, fled step-dame study's blows; 10
And others' feet still seemed but strangers in my way.
Thus great with child to speak, and helpless in my throes,
  Biting my truant pen, beating myself for spite,
  'Fool', said my muse to me; 'look in thy heart, and write'.

2

Not at first sight, nor with a dribbed shot,
  Love gave the wound which while I breathe will bleed:
  But known worth did in mine of time proceed,
Till by degrees it had full conquest got.
I saw, and liked; I liked, but loved not;
  I loved, but straight did not what love decreed:
  At length to love's decrees I, forced, agreed,
Yet with repining at so partial lot.
  Now even that footstep of lost liberty
Is gone, and now like slave-born Muscovite 10
I call it praise to suffer tyranny;
And now employ the remnant of my wit
  To make myself believe that all is well,
  While with a feeling skill I paint my hell.

3

Let dainty wits cry on the sisters nine,
That bravely masked, their fancies may be told:

Or Pindar's apes, flaunt they in phrases fine,
Enam'lling with pied flowers their thoughts of gold:
  Or else let them in statelier glory shine,
Ennobling new-found tropes with problems old:
Or with strange similes enrich each line,
Of herbs or beasts, which Ind or Afric hold.
  For me, in sooth, no muse but one I know;
  Phrases and problems from my reach do grow, 10
And strange things cost too dear for my poor sprites.
  How then? Even thus: in Stella's face I read
  What love and beauty be; then all my deed
But copying is, what in her nature writes.

4

Virtue, alas, now let me take some rest:
Thou sett'st a bate between my will and wit.
If vain love have my simple soul oppressed,
Leave what thou lik'st not, deal not thou with it.
  Thy sceptre use in some old Cato's breast;
Churches or schools are for thy seat more fit.
I do confess—pardon a fault confessed—
My mouth too tender is for thy hard bit.
  But if that needs thou wilt usurping be
  The little reason that is left in me, 10
And still the effect of thy persuasions prove:
  I swear, my heart such one shall show to thee
  That shrines in flesh so true a deity,
That, virtue, thou thy self shalt be in love.

5

It is most true, that eyes are formed to serve
The inward light; and that the heavenly part
Ought to be king, from whose rules who do swerve,
Rebels to Nature, strive for their own smart.
  It is most true, what we call Cupid's dart,
An image is, which for ourselves we carve;
And, fools, adore in temple of our heart,
Till that good god make church and churchman starve.
  True, that true beauty virtue is indeed,
Whereof this beauty can be but a shade, 10

Which elements with mortal mixture breed;
True, that on earth we are but pilgrims made,
And should in soul up to our country move;
True; and yet true, that I must Stella love.

6

Some lovers speak, when they their muses entertain,
Of hopes begot by fear, of wot not what desires,
Of force of heavenly beams, infusing hellish pain,
Of living deaths, dear wounds, fair storms and freezing fires.
Some one his song in Jove, and Jove's strange tales, attires,
Broidered with bulls and swans, powdered with golden rain.
Another, humbler, wit to shepherd's pipe retires,
Yet hiding royal blood full often in rural vein.
To some a sweetest plaint a sweetest style affords,
While tears pour out his ink, and sighs breathe out his words, 10
His paper, pale despair, and pain his pen doth move.
I can speak what I feel, and feel as much as they,
But think that all the map of my state I display,
When trembling voice brings forth, that I do Stella love.

7

When nature made her chief work, Stella's eyes,
In colour black why wrapped she beams so bright?
Would she in beamy black, like painter wise,
Frame daintiest lustre, mixed of shades and light?
Or did she else that sober hue devise
In object best to knit and strength our sight,
Lest, if no veil those brave gleams did disguise,
They, sun-like, should more dazzle than delight?
Or would she her miraculous power show,
That, whereas black seems beauty's contrary, 10
She even in black doth make all beauties flow?
Both so, and thus: she minding love should be
Placed ever there, gave him this mourning weed
To honour all their deaths, who for her bleed.

8

Love, born in Greece, of late fled from his native place,
Forced by a tedious proof, that Turkish hardened heart

Is no fit mark to pierce with his fine pointed dart;
And pleased with our soft peace, stayed here his flying race.
But finding these North climes too coldly him embrace,
Not used to frozen clips, he strave to find some part
Where with most ease and warmth he might employ his art.
At length he perched himself in Stella's joyful face,
Whose fair skin, beamy eyes, like morning sun on snow,
Deceived the quaking boy, who thought from so pure light 10
Effects of lively heat must needs in nature grow.
But she, most fair, most cold, made him thence take his flight
To my close heart, where, while some firebrands he did lay,
He burnt unwares his wings, and cannot fly away.

9

Queen Virtue's court, which some call Stella's face,
Prepared by Nature's chiefest furniture,
Hath his front built of alablaster pure;
Gold is the covering of that stately place.
The door, by which sometimes comes forth her grace,
Red porphyr is, which lock of pearl makes sure;
Whose porches rich (which name of 'cheeks' endure)
Marble, mixed red and white, do interlace.
The windows now through which this heavenly guest
Looks o'er the world, and can find nothing such 10
Which dare claim from those lights the name of 'best',
Of touch they are that without touch doth touch,
Which Cupid's self from Beauty's mine did draw:
Of touch they are, and poor I am their straw.

10

Reason, in faith thou art well served, that still
Would'st brabbling be with sense and love in me.
I rather wished thee climb the muses' hill,
Or reach the fruit of nature's choicest tree,
Or seek heaven's course, or heaven's inside, to see.
Why should'st thou toil our thorny soil to till?
Leave sense, and those which sense's objects be:
Deal thou with powers of thoughts, leave love to will.
But thou would'st needs fight both with love and sense,
With sword of wit giving wounds of dispraise, 10
Till downright blows did foil thy cunning fence:

For soon as they strake thee with Stella's rays,
Reason, thou kneeled'st, and offered'st straight to prove
By reason good, good reason her to love.

11

In truth, O Love, with what a boyish kind
Thou dost proceed in thy most serious ways:
That when the heaven to thee his best displays
Yet of that best thou leav'st the best behind.
For like a child, that some fair book doth find,
With gilded leaves or coloured vellum plays,
Or at the most, on some fine picture stays,
But never heeds the fruit of writer's mind:
So when thou saw'st, in nature's cabinet,
Stella, thou straight look'st babies in her eyes, 10
In her cheek's pit thou did'st thy pit-fold set,
And in her breast bo-peep or couching lies,
Playing and shining in each outward part:
But, fool, seek'st not to get into her heart.

12

Cupid, because thou shin'st in Stella's eyes,
That from her locks, thy day-nets, nonc 'scapes free,
That those lips swell, so full of thee they be,
That her sweet breath makes oft thy flames to rise,
That in her breast thy pap well sugared lies,
That her grace gracious makes thy wrongs, that she,
What words so e'er she speaks, persuades for thee,
That her clear voice lifts thy fame to the skies;
Thou countest Stella thine, like those whose powers,
Having got up a breach by fighting well, 10
Cry, 'Victory, this fair day all is ours!'
O no, her heart is such a citadel,
So fortified with wit, stored with disdain,
That to win it, is all the skill and pain.

13

Phoebus was judge between Jove, Mars, and Love,
Of those three gods, whose arms the fairest were.
Jove's golden shield did eagle sables bear,

Whose talents held young Ganymede above:
But in vert field Mars bare a golden spear
  Which through a bleeding heart his point did shove.
  Each had his crest: Mars carried Venus' glove,
Jove on his helm the thunderbolt did rear.
Cupid then smiles, for on his crest there lies
  Stella's fair hair, her face he makes his shield, 10
  Where roses gules are borne in silver field.
Phoebus drew wide the curtains of the skies
  To blaze these last, and sware devoutly then,
  The first, thus matched, were scarcely gentlemen.

14

Alas, have I not pain enough, my friend,
  Upon whose breast a fiercer gripe doth tire
  Than did on him who first stale down the fire,
While Love on me doth all his quiver spend,
But with your rhubarb words you must contend
  To grieve me worse, in saying that desire
  Doth plunge my well-formed soul even in the mire
Of sinful thoughts, which do in ruin end?
  If that be sin, which doth the manners frame,
Well stayed with truth in word, and faith of deed, 10
Ready of wit, and fearing nought but shame:
If that be sin, which in fixed hearts doth breed
  A loathing of all loose unchastity:
  Then love is sin, and let me sinful be.

15

You that do search for every purling spring
  Which from the ribs of old Parnassus flows;
  And every flower, not sweet perhaps, which grows
Near thereabouts, into your poesy wring;
You that do dictionary's method bring
  Into your rhymes, running in rattling rows;
  You that poor Petrarch's long-deceased woes
With new-born sighs and denizened wit do sing:
  You take wrong ways, those far-fet helps be such
  As do bewray a want of inward touch: 10
And sure at length stol'n goods do come to light.

But if (both for your love and skill) your name
You seek to nurse at fullest breasts of fame,
Stella behold, and then begin to endite.

16

In nature apt to like, when I did see,
Beauties, which were of many carats fine,
My boiling sprites did thither soon incline,
And, love, I thought that I was full of thee.
But finding not those restless flames in me
Which others said did make their souls to pine,
I thought those babes of some pin's hurt did whine,
By my love judging what love's pain might be.
But while I thus with this young lion played,
Mine eyes (shall I say cursed or blessed?) beheld 10
Stella: now she is named, need more be said?
In her sight I a lesson new have spelled;
I now have learned love right, and learned even so
As who by being poisoned doth poison know.

17

His mother dear Cupid offended late,
Because that Mars, grown slacker in her love,
With pricking shot he did not throughly move,
To keep the pace of their first loving state.
The boy refused, for fear of Mars's hate,
Who threatened stripes if he his wrath did prove.
But she in chafe him from her lap did shove,
Brake bow, brake shafts, while Cupid weeping sate:
Till that his grandame, Nature, pitying it,
Of Stella's brows made him two better bows, 10
And in her eyes of arrows infinite.
O how for joy he leaps, O how he crows,
And straight therewith, like wags new got to play,
Falls to shrewd turns; and I was in his way.

18

With what sharp checks I in myself am shent
When into reason's audit I do go,
And by just counts myself a bankrupt know

Of all those goods, which heaven to me hath lent,
Unable quit to pay even nature's rent,
Which unto it by birthright I do owe:
And which is worse, no good excuse can show,
But that my wealth I have most idly spent.
My youth doth waste, my knowledge brings forth toys,
My wit doth strive those passions to defend 10
Which for reward spoil it with vain annoys.
I see my course to lose myself doth bend:
I see, and yet no greater sorrow take
Than that I lose no more for Stella's sake.

19

On Cupid's bow how are my heart-strings bent,
That see my wrack, and yet embrace the same!
When most I glory, then I feel most shame:
I willing run, yet while I run, repent.
My best wits still their own disgrace invent;
My very ink turns straight to Stella's name;
And yet my words, as them my pen doth frame,
Avise themselves that they are vainly spent.
Fthough she pass all things, yet what is all
That unto me, who fare like him that both 10
Looks to the skies, and in a ditch doth fall?
O let me prop my mind, yet in his growth,
And not in nature for best fruits unfit.
'Scholar', saith Love, 'bend hitherward your wit.'

20

Fly, fly, my friends, I have my death wound, fly;
See there that boy, that murth'ring boy I say,
Who like a thief hid in dark bush doth lie,
Till bloody bullet get him wrongful prey.
So tyrant he no fitter place could spy,
Nor so fair level in so secret stay
As that sweet black which veils the heavn'ly eye;
There himself with his shot he close doth lay.
Poor passenger, pass now thereby I did,
And stayed, pleased with the prospect of the place, 10
While that black hue from me the bad guest hid:

But straight I saw motions of lightning grace,
  And then descried the glist'ring of his dart:
  But ere I could fly thence, it pierced my heart.

21

Your words, my friend, right healthful caustics, blame
  My young mind marred, whom love doth windlass so
  That mine own writings like bad servants show,
My wits, quick in vain thoughts, in virtue lame;
That Plato I read for nought, but if he tame
  Such coltish gyres; that to my birth I owe
  Nobler desires, lest else that friendly foe,
  Great expectation, wear a train of shame.
  For since mad March great promise made of me,
If now the May of my years much decline, 10
What can be hoped my harvest time will be?
Sure you say well; your wisdom's golden mine
  Dig deep with learning's spade; now tell me this,
Hath this world aught so fair as Stella is?

22

In highest way of heaven the sun did ride,
  Progressing then from fair twins' golden place,
  Having no scarf of clouds before his face,
But shining forth of heat in his chief pride,
When some fair ladies, by hard promise tied,
  On horseback met him in his furious race;
  Yet each prepared, with fan's well-shading grace,
From that foe's wounds their tender skins to hide.
Stella alone with face unarmed marched,
  Either to do like him, which open shone, 10
  Or careless of the wealth because her own;
Yet were the hid and meaner beauties parched,
  Her daintiest bare went free. The cause was this:
  The sun, which others burned, did her but kiss.

23

The curious wits, seeing dull pensiveness
  Bewray itself in my long settled eyes,
  Whence those same fumes of melancholy rise

With idle pains, and missing aim, do guess.
Some, that know how my spring I did address,
  Deem that my muse some fruit of knowledge plies;
  Others, because the prince my service tries,
Think that I think state errors to redress.
  But harder judges judge ambition's rage,
Scourge of itself, still climbing slippery place, 10
Holds my young brain captived in golden cage.
O fools, or over-wise: alas, the race
  Of all my thoughts hath neither stop nor start
  But only Stella's eyes and Stella's heart.

24

Rich fools there be, whose base and filthy heart
Lies hatching still the goods wherein they flow;
And damning their own selves to Tantal's smart,
Wealth breeding want, more blessed, more wretched grow.
  Yet to those fools heaven such wit doth impart
As what their hands do hold, their heads do know,
And knowing, love, and loving, lay apart,
As sacred things, far from all danger's show.
  But that rich fool, who by blind fortune's lot
The richest gem of love and life enjoys, 10
And can with foul abuse such beauties blot,
Let him, deprived of sweet but unfelt joys,
  Exiled for aye from those high treasures which
  He knows not, grow in only folly rich.

25

The wisest scholar of the wight most wise,
By Phoebus' doom, with sugared sentence says,
That virtue, if it once met with our eyes,
Strange flames of love it in our souls would raise;
  But for that man with pain this truth descries,
While he each thing in sense's balance weighs,
And so nor will, nor can, behold those skies
Which inward sun to heroic mind displays:
  Virtue of late, with virtuous care to stir
Love of herself, takes Stella's shape, that she 10
To mortal eyes might sweetly shine in her.

It is most true, for since I her did see,
  Virtue's great beauty in that face I prove,
  And find the effect, for I do burn in love.

26

Though dusty wits dare scorn astrology,
And fools can think those lamps of purest light,
Whose numbers, ways, greatness, eternity,
Promising wonders, wonder to invite,
  To have for no cause birthright in the sky,
But for to spangle the black weeds of night;
Or for some brawl, which in that chamber high
They should still dance, to please a gazer's sight:
  For me, I do Nature unidle know,
And know great causes great effects procure, 10
And know those bodies high reign on the low.
And if these rules did fail, proof makes me sure,
  Who oft fore-judge my after-following race
  By only those two eyes in Stella's face.

27

Because I oft, in dark abstracted guise,
  Seem most alone in greatest company,
  With dearth of words, or answers quite awry,
To them that would make speech of speech arise,
They deem, and of their doom the rumour flies,
  That poison foul of bubbling pride doth lie
  So in my swelling breast, that only I
Fawn on myself, and others do despise.
  Yet pride, I think, doth not my soul possess,
Which looks too oft in his unflatt'ring glass; 10
But one worse fault, ambition, I confess,
That makes me oft my best friends overpass,
  Unseen, unheard, while thought to highest place
  Bends all his powers, even unto Stella's grace.

28

You that with allegory's curious frame
  Of others' children changelings use to make,
  With me those pains, for God's sake, do not take;

I list not dig so deep for brazen fame.
When I say 'Stella', I do mean the same
  Princess of beauty, for whose only sake
  The reins of love I love, though never slake,
And joy therein, though nations count it shame.
  I beg no subject to use eloquence,
Nor in hid ways to guide philosophy. 10
Look at my hands for no such quintessence,
But know that I, in pure simplicity,
  Breathe out the flames which burn within my heart,
  Love only reading unto me this art.

29

Like some weak lords, neighboured by mighty kings,
  To keep themselves and their chief cities free,
  Do easily yield, that all their coasts may be
Ready to store their camps of needful things:
So Stella's heart, finding what power love brings,
  To keep itself in life and liberty,
  Doth willing grant, that in the frontiers he
Use all to help his other conquerings.
And thus her heart escapes; but thus her eyes
  Serve him with shot, her lips his heralds are, 10
  Her breasts his tents, legs his triumphal car,
Her flesh his food, her skin his armour brave;
And I, but for because my prospect lies
Upon that coast, am giv'n up for a slave.

30

Whether the Turkish new moon minded be
  To fill his horns this year on Christian coast;
  How Pole's right king means, without leave of host,
To warm with ill-made fire cold Muscovy;
If French can yet three parts in one agree;
  What now the Dutch in their full diets boast;
  How Holland hearts, now so good towns be lost,
Trust in the pleasing shade of Orange tree;
  How Ulster likes of that same golden bit
Wherewith my father once made it half tame; 10
If in the Scottish court be welt'ring yet;

These questions busy wits to me do frame.
I, cumbered with good manners, answer do,
But know not how, for still I think of you.

31

With how sad steps, O moon, thou climb'st the skies;
How silently, and with how wan a face.
What, may it be that even in heav'nly place
That busy archer his sharp arrows tries?
Sure, if that long-with-love-acquainted eyes
Can judge of love, thou feel'st a lover's case;
I read it in thy looks; thy languished grace
To me, that feel the like, thy state descries.
Then even of fellowship, O moon, tell me,
Is constant love deemed there but want of wit? 10
Are beauties there as proud as here they be?
Do they above love to be loved, and yet
Those lovers scorn whom that love doth possess?
Do they call virtue there ungratefulness?

32

Morpheus, the lively son of deadly sleep,
Witness of life to them that living die;
A prophet oft, and oft an history,
A poet eke, as humours fly or creep;
Since thou in me so sure a power dost keep
That never I with closed-up sense do lie
But by thy work my Stella I descry
Teaching blind eyes both how to smile and weep,
Vouchsafe of all acquaintance this to tell:
Whence hast thou ivory, rubies, pearl and gold 10
To show her skin, lips, teeth and head so well?
'Fool,' answers he; 'no Ind's such treasures hold,
But from thy heart, while my sire charmeth thee,
Sweet Stella's image I do steal to me.'

33

I might (unhappy word), O me, I might,
And then would not, or could not, see my bliss:
Till now, wrapped in a most infernal night,

I find how heavenly day, wretch, I did miss.
  Heart, rend thyself, thou dost thyself but right;
No lovely Paris made thy Helen his;
No force, no fraud, robbed thee of thy delight;
Nor Fortune of thy fortune author is;
  But to myself myself did give the blow,
While too much wit (forsooth) so troubled me 10
That I respects for both our sakes must show:
And yet could not by rising morn foresee
  How fair a day was near. O punished eyes
  That I had been more foolish, or more wise!

34

Come, let me write. 'And to what end?' To ease
  A burdened heart. 'How can words ease, which are
  The glasses of thy daily vexing care?'
Oft cruel fights well pictured forth do please.
'Art not ashamed to publish thy disease?'
  Nay, that may breed my fame, it is so rare.
  'But will not wise men think thy words fond ware?'
Then be they close, and so none shall displease.
  'What idler thing, than speak and not be heard?'
What harder thing than smart, and not to speak? 10
Peace, foolish wit; with wit my wit is marred.
Thus write I while I doubt to write, and wreak
  My harms on ink's poor loss; perhaps some find
  Stella's great powers, that so confuse my mind.

35

What may words say, or what may words not say,
Where truth itself must speak like flattery?
Within what bounds can one his liking stay,
Where nature doth with infinite agree?
  What Nestor's counsels can my flames allay,
Since reason's self doth blow the coal in me?
And ah, what hope that hope should once see day,
Where Cupid is sworn page to chastity?
Honour is honoured, that thou dost possess
  Him as thy slave; and now long needy fame 10
  Doth even grow rich, naming my Stella's name.

Wit learns in thee perfection to express;
Not thou by praise, but praise in thee is raised;
It is a praise to praise, when thou art praised.

36

Stella, whence doth this new assault arise,
A conquered, yelden, ransacked heart to win?
Whereto long since, through my long battered eyes,
Whole armies of thy beauties entered in;
And there, long since, love, thy lieutenant lies;
My forces razed, thy banners raised within.
Of conquest do not these effects suffice,
But wilt new war upon thine own begin?
With so sweet voice, and by sweet nature so,
In sweetest strength, so sweetly skilled withal, 10
In all sweet stratagems sweet art can show,
That not my soul, which at thy foot did fall,
Long since forced by thy beams, but stone nor tree,
By sense's privilege, can 'scape from thee.

37

My mouth doth water, and my breast doth swell,
My tongue doth itch, my thoughts in labour be;
Listen then, lordings, with good ear to me,
For of my life I must a riddle tell.
Towards Aurora's court a nymph doth dwell,
Rich in all beauties which man's eye can see;
Beauties so far from reach of words, that we
Abase her praise, saying she doth excel;
Rich in the treasure of deserved renown;
Rich in the riches of a royal heart; 10
Rich in those gifts which give the eternal crown;
Who though most rich in these, and every part
Which make the patents of true worldly bliss,
Hath no misfortune, but that Rich she is.

38

This night, while sleep begins with heavy wings
To hatch mine eyes, and that unbitted thought
Doth fall to stray, and my chief powers are brought

To leave the sceptre of all subject things,
The first that straight my fancy's error brings
  Unto my mind, is Stella's image, wrought
  By love's own self; but with so curious draught
That she, methinks, not only shines, but sings.
  I start, look, heark; but what in closed-up sense
Was held, in opened sense it flies away, 10
Leaving me nought but wailing eloquence.
I, seeing better sights in sight's decay,
  Called it anew, and wooed sleep again:
  But him, her host, that unkind guest had slain.

39

Come sleep, O sleep, the certain knot of peace,
The baiting place of wit, the balm of woe,
The poor man's wealth, the prisoner's release,
The indifferent judge between the high and low;
  With shield of proof shield me from out the press
Of those fierce darts despair at me doth throw:
O make in me those civil wars to cease;
I will good tribute pay, if thou do so.
  Take thou of me smooth pillows, sweetest bed,
A chamber deaf to noise, and blind to light; 10
A rosy garland, and a weary head;
And if these things, as being thine by right,
  Move not thy heavy grace, thou shalt in me,
  Livelier than elsewhere, Stella's image see.

40

As good to write, as for to lie and groan.
  O Stella dear, how much thy power hath wrought,
  That hast my mind, none of the basest, brought
My still kept course, while others sleep, to moan.
Alas, if from the height of virtue's throne
  Thou canst vouchsafe the influence of a thought
  Upon a wretch, that long thy grace hath sought;
Weigh then how I by thee am overthrown:
  And then, think thus: although thy beauty be
  Made manifest by such a victory, 10
Yet noblest conquerers do wrecks avoid.

Since then thou hast so far subdued me,
That in my heart I offer still to thee,
O, do not let thy temple be destroyed.

41

Having this day my horse, my hand, my lance,
Guided so well, that I obtained the prize,
Both by the judgement of the English eyes
And of some sent from that sweet enemy, France;
Horsemen my skill in horsemanship advance;
Town-folks my strength; a daintier judge applies
His praise to sleight, which from good use doth rise;
Some lucky wits impute it but to chance;
Others, because of both sides I do take
My blood from them, who did excel in this, 10
Think nature me a man of arms did make.
How far they shoot awry! The true cause is,
Stella looked on, and from her heavenly face
Sent forth the beams, which made so fair my race.

42

O eyes, which do the spheres of beauty move,
Whose beams be joys, whose joys all virtues be,
Who, while they make love conquer, conquer love;
The schools where Venus hath learned chastity;
O eyes, where humble looks most glorious prove,
Only loved tyrants, just in cruelty;
Do not, O do not, from poor me remove;
Keep still my zenith, ever shine on me.
For though I never see them, but straight ways
My life forgets to nourish languished sprites; 10
Yet still on me, O eyes, dart down your rays;
And if from majesty of sacred lights,
Oppressing mortal sense, my death proceed,
Wracks triumphs be, which love (high set) doth breed.

43

Fair eyes, sweet lips, dear heart, that foolish I
Could hope by Cupid's help on you to prey;
Since to himself he doth your gifts apply.

And his main force, choice sport, and easeful stay.
  For when he will see who dare him gainsay,
Then with those eyes he looks; lo, by and by
Each soul doth at love's feet his weapons lay,
Glad if for her he give them leave to die.
  When he will play, then in her lips he is,
Where, blushing red, that love's self doth them love, 10
With either lip he doth the other kiss
But when he will for quiet's sake remove
  From all the world, her heart is then his room,
  Where well he knows, no man to him can come.

44

My words, I know, do well set forth my mind;
  My mind bemoans his sense of inward smart;
  Such smart may pity claim of any heart;
Her heart (sweet heart) is of no tiger's kind:
And yet she hears, yet I no pity find,
  But more I cry, less grace she doth impart.
  Alas, what cause is there so overthwart,
That nobleness itself makes thus unkind?
  I much do guess, yet find no truth save this:
That when the breath of my complaints doth touch 10
Those dainty doors unto the court of bliss,
The heavenly nature of that place is such
  That once come there, the sobs of mine annoys
  Are metamorphosed straight to tunes of joys.

45

Stella oft sees the very face of woe
  Painted in my beclouded stormy face;
  But cannot skill to pity my disgrace,
Not though thereof the cause herself she know;
Yet hearing late a fable, which did show
  Of lovers never known a grievous case,
  Pity thereof gat in her breast such place
That, from that sea derived, tears' spring did flow.
  Alas, if fancy drawn by imaged things,
Though false, yet with free scope more grace doth breed 10
Than servant's wrack, where new doubts honour brings;

Then think, my dear, that you in me do read
Of lover's ruin some sad tragedy:
I am not I, pity the tale of me.

46

I cursed thee oft; I pity now thy case,
Blind-hitting boy, since she that thee and me
Rules with a beck, so tyrannizeth thee,
That thou must want or food, or dwelling-place.
For she protests to banish thee her face—
Her face? O love, a rogue thou then should'st be
If love learn not alone to love and see,
Without desire to feed of further grace.
Alas poor wag, that now a scholar art
To such a school-mistress, whose lessons new 10
Thou needs must miss, and so thou needs must smart.
Yet dear, let me this pardon get of you,
So long (though he from book mich to desire)
Till without fuel you can make hot fire.

47

What, have I thus betrayed my liberty?
Can those black beams such burning marks engrave
In my free side? or am I born a slave,
Whose neck becomes such yoke of tyranny?
Or want I sense to feel my misery?
Or spirit, disdain of such disdain to have,
Who for long faith, though daily help I crave,
May get no alms, but scorn of beggary?
Virtue, awake: beauty but beauty is;
I may, I must, I can, I will, I do 10
Leave following that, which it is gain to miss.
Let her go. Soft, but here she comes. Go to,
Unkind, I love you not—: O me, that eye
Doth make my heart give to my tongue the lie.

48

Soul's joy, bend not those morning stars from me,
Where virtue is made strong by beauty's might,
Where love is chasteness, pain doth learn delight,

And humbleness grows one with majesty.
Whatever may ensue, O let me be
  Co-partner of the riches of that sight;
  Let not mine eyes be hell-driven from that light;
O look, O shine, O let me die, and see.
  For though I oft my self of them bemoan,
  That through my heart their beamy darts be gone, 10
Whose cureless wounds even now most freshly bleed;
  Yet since my death-wound is already got,
  Dear killer, spare not thy sweet cruel shot:
A kind of grace it is to slay with speed.

49

I on my horse, and love on me, doth try
  Our horsemanships, while by strange work I prove
  A horseman to my horse, a horse to love;
And now man's wrongs in me, poor beast, descry.
The reins wherewith my rider doth me tie
  Are humbled thoughts, which bit of reverence move,
  Curbed in with fear, but with gilt boss above
Of hope, which makes it seem fair to the eye.
  The wand is will; thou, fancy, saddle art,
Girt fast by memory; and while I spur 10
My horse, he spurs with sharp desire my heart;
He sits me fast, however I do stir;
  And now hath made me to his hand so right
  That in the manage myself takes delight.

50

Stella, the fullness of my thoughts of thee
Cannot be stayed within my panting breast,
But they do swell and struggle forth of me,
Till that in words thy figure be expressed.
  And yet, as soon as they so formed be,
According to my lord love's own behest,
With sad eyes I their weak proportion see,
To portrait that which in this world is best;
  So that I cannot choose but write my mind,
And cannot choose but put out what I write, 10
While those poor babes their death in birth do find:

And now my pen these lines had dashed quite,
But that they stopped his fury from the same,
Because their forefront bare sweet Stella's name.

51

Pardon, mine ears, both I and they do pray,
So may your tongue still fluently proceed,
To them that do such entertainment need,
So may you still have somewhat new to say.
On silly me do not the burden lay
Of all the grave conceits your brain doth breed;
But find some Hercules to bear, in steed
Of Atlas tired, your wisdom's heavenly sway.
For me, while you discourse of courtly tides,
Of cunning'st fishers in most troubled streams, 10
Of straying ways, when valiant error guides;
Meanwhile my heart confers with Stella's beams,
And is even irked that so sweet comedy
By such unsuited speech should hindered be.

52

A strife is grown between virtue and love,
While each pretends that Stella must be his.
Her eyes, her lips, her all, saith love, do this,
Since they do wear his badge, most firmly prove.
But virtue thus that title doth disprove:
That Stella (O dear name) that Stella is
That virtuous soul, sure heir of heavenly bliss,
Not this fair outside, which our hearts doth move;
And therefore, though her beauty and her grace
Be love's indeed, in Stella's self he may 10
By no pretence claim any manner place.
Well, love, since this demur our suit doth stay,
Let virtue have that Stella's self; yet thus,
That virtue but that body grant to us.

53

In martial sports I had my cunning tried,
And yet to break more staves did me address,
While with the people's shouts, I must confess,

Youth, luck and praise even filled my veins with pride;
When Cupid, having me, his slave, descried
In Mars's livery, prancing in the press:
'What now, sir fool,' said he; 'I would no less,
Look here, I say.' I looked, and Stella spied,
Who hard by made a window send forth light.
My heart then quaked, then dazzled were mine eyes, 10
One hand forgot to rule, th'other to fight;
Nor trumpet's sound I heard, nor friendly cries;
My foe came on, and beat the air for me,
Till that her blush taught me my shame to see.

54

Because I breathe not love to every one,
Nor do not use set colours for to wear,
Nor nourish special locks of vowed hair,
Nor give each speech a full point of a groan,
The courtly nymphs, acquainted with the moan
Of them, who in their lips love's standard bear:
'What, he?' say they of me, 'now I dare swear,
He cannot love; no, no, let him alone.'
And think so still, so Stella know my mind.
Profess indeed I do not Cupid's art; 10
But you fair maids, at length this true shall find,
That his right badge is but worn in the heart;
Dumb swans, not chattering pies, do lovers prove;
They love indeed, who quake to say they love.

55

Muses, I oft invoked your holy aid,
With choicest flowers my speech to engarland so
That it, despised in true but naked show,
Might win some grace in your sweet skill arrayed;
And oft whole troops of saddest words I stayed,
Striving abroad a-foraging to go,
Until by your inspiring I might know
How their black banner might be best displayed.
But now I mean no more your help to try,
Nor other sugaring of my speech to prove, 10
But on her name incessantly to cry:

For let me but name her, whom I do love,
So sweet sounds straight mine ear and heart do hit
That I well find no eloquencc like it.

56

Fie, school of patience, fie; your lesson is
Far, far too long to learn it without book:
What, a whole week without one piece of look,
And think I should not your large precepts miss?
When I might read those letters fair of bliss,
Which in her face teach virtue, I could brook
Somewhat thy leaden counsels, which I took
As of a friend that meant not much amiss:
But now that I, alas, do want her sight,
What, dost thou think that I can ever take 10
In thy cold stuff a phlegmatique delight?
No, patience; if thou wilt my good, then make
Her come, and hear with patience my desire,
And then with patience bid me bear my fire.

57

Woe, having made with many fights his own
Each sense of mine, each gift, each power of mind,
Grown now his slaves, he forced them out to find
The thorough'st words, fit for woe's self to groan,
Hoping that when they might find Stella alone,
Before she could prepare to be unkind,
Her soul, armed but with such a dainty rind,
Should soon be pierced with sharpness of the moan.
She heard my plaints, and did not only hear,
But them (so sweet she is) most sweetly sing, 10
With that fair breast making woe's darkness clear.
A pretty case! I hoped her to bring
To feel my griefs, and she with face and voice
So sweets my pains, that my pains me rejoice.

58

Doubt there hath been, when with his golden chain
The orator so far men's hearts doth bind
That no pace else their guided steps can find

But as he them more short or slack doth rein
Whether with words this sovereignty he gain
  Clothed with fine tropes, with strongest reasons lined,
  Or else pronouncing grace, wherewith his mind
Prints his own lively form in rudest brain.
  Now judge by this: in piercing phrases late
  The anatomy of all my woes I wrate, 10
Stella's sweet breath the same to me did read.
  O voice, O face, maugre my speech's might,
Which wooed woe, most ravishing delight
Even those sad words even in sad me did breed.

59

Dear, why make you more of a dog than me?
  If he do love, I burn, I burn in love;
  If he wait well, I never thence would move;
If he be fair, yet but a dog can be.
Little he is, so little worth is he;
  He barks, my songs thy own voice oft doth prove;
  Bidden, perhaps he fetcheth thee a glove;
But I unbid fetch even my soul to thee.
  Yet while I languish, him that bosom clips,
That lap doth lap, nay lets, in spite of spite, 10
This sour-breathed mate taste of those sugared lips.
Alas, if you grant only such delight
  To witless things, then love, I hope (since wit
  Becomes a clog) will soon ease me of it.

60

When my good angel guides me to the place
  Where all my good I do in Stella see,
  That heaven of joys throws only down on me
Thundered disdains, and lightnings of disgrace;
But when the rugged'st step of fortune's race
  Makes me fall from her sight, then sweetly she
  With words, wherein the muses' treasures be,
Shows love and pity to my absent case.
  Now I, wit-beaten long by hardest fate,
So dull am, that I cannot look into 10
The ground of this fierce love and lovely hate;

Then some good body tell me how I do,
Whose presence absence, absence presence is;
Blessed in my curse, and cursed in my bliss.

61

Oft with true sighs, oft with uncalled tears,
Now with slow words, now with dumb eloquence,
I Stella's eyes assail, invade her ears;
But this at last is her sweet-breathed defence:
That who indeed infelt affection bears,
So captives to his saint both soul and sense
That wholly hers, all selfness he forbears;
Thence his desires he learns, his life's course thence.
Now since her chaste mind hates this love in me,
With chastened mind I straight must show that she 10
Shall quickly me from what she hates remove.
O doctor Cupid, thou for me reply;
Driven else to grant, by angel's sophistry,
That I love not, without I leave to love.

62

Late tired with woe, even ready for to pine
With rage of love, I called my love unkind;
She in whose eyes love, though unfelt, doth shine,
Sweet said that I true love in her should find.
I joyed, but straight thus watered was my wine,
That love she did, but loved a love not blind,
Which would not let me, whom she loved, decline
From nobler course, fit for my birth and mind:
And therefore, by her love's authority,
Willed me these tempests of vain love to fly, 10
And anchor fast myself on virtue's shore.
Alas, if this the only metal be
Of love, new-coined to help my beggary,
Dear, love me not, that you may love me more.

63

O grammar rules, O now your virtues show:
So children still read you with awful eyes,
As my young dove may in your precepts wise,

Her grant to me, by her own virtue, know.
For late, with heart most high, with eyes most low,
  I craved the thing, which ever she denies:
  She, lightning love, displaying Venus' skies,
Lest once should not be heard, twice said, 'No, no.'
  Sing then, my muse, now Io Paean sing;
  Heavens, envy not at my high triumphing, 10
But grammar's force with sweet success confirm.
  For grammar says (O this, dear Stella, weigh),
  For grammar says (to grammar who says nay?)
That in one speech two negatives affirm.

First song

Doubt you to whom my muse these notes intendeth,
Which now my breast, o'ercharged, to music lendeth?
To you, to you, all song of praise is due;
Only in you my song begins and endeth.

Who hath the eyes which marry state with pleasure,
Who keeps the key of nature's chiefest treasure?
To you, to you, all song of praise is due;
Only for you the heaven forgat all measure.

Who hath the lips, where wit in fairness reigneth,
Who womankind at once both decks and staineth? 10
To you, to you, all song of praise is due;
Only by you Cupid his crown maintaineth.

Who hath the feet, whose step all sweetness planteth,
Who else for whom fame worthy trumpets wanteth?
To you, to you, all song of praise is due;
Only to you her sceptre Venus granteth.

Who hath the breast, whose milk doth passions nourish,
Whose grace is such, that when it chides doth cherish?
To you, to you, all song of praise is due;
Only through you the tree of life doth flourish. 20

Who hath the hand which without stroke subdueth,
Who long-dead beauty with increase reneweth?
To you, to you, all song of praise is due;
Only at you all envy hopeless rueth.

Who hath the hair which loosest, fastest, tieth?
Who makes a man live then glad, when he dieth?
To you, to you, all song of praise is due;
Only of you the flatterer never lieth.

Who hath the voice which soul from senses sunders?
Whose force but yours the bolts of beauty thunders? 30
To you, to you, all song of praise is due;
Only with you not miracles are wonders.

Doubt you to whom my muse these notes intendeth,
Which now my breast, o'ercharged, to music lendeth?
To you, to you, all song of praise is due;
Only in you my song begins and endeth.

64

No more, my dear, no more these counsels try;
  O give my passions leave to run their race.
  Let fortune lay on me her worst disgrace,
Let folk o'ercharged with brain against me cry,
Let clouds bedim my face, break in mine eye,
  Let me no steps but of lost labour trace,
  Let all the earth with scorn recount my case,
But do not will me from my love to fly.
  I do not envy Aristotle's wit,
Nor do aspire to Caesar's bleeding fame, 10
Nor aught do care, though some above me sit,
Nor hope, nor wish, another course to frame,
  But that which once may win thy cruel heart:
  Thou art my wit, and thou my virtue art.

65

Love, by sure proof I may call thee unkind,
That giv'st no better ear to my just cries;
Thou whom to me such my good turns should bind,

As I may well recount, but none can prize.
For when, nak'd boy, thou could'st no harbour find
In this old world, grown now so too too wise,
I lodged thee in my heart; and being blind
By nature born, I gave to thee mine eyes.
Mine eyes, my light, my heart, my life, alas,
If so great services may scorned be, 10
Yet let this thought thy tigerish courage pass,
That I, perhaps, am somewhat kin to thee:
Since in thine arms, if learn'd fame truth hath spread,
Thou bear'st the arrow, I the arrow head.

66

And do I see some cause a hope to feed,
Or doth the tedious burden of long woe
In weakened minds, quick apprehending breed,
Of every image, which may comfort show?
I cannot brag of word, much less of deed;
Fortune wheels still with me in one sort slow;
My wealth no more, and no whit less my need;
Desire still on the stilts of fear doth go.
And yet amid all fears, a hope there is
Stol'n to my heart, since last fair night; nay day: 10
Stella's eyes sent to me the beams of bliss,
Looking on me, while I looked other way;
But when mine eyes back to their heaven did move,
They fled with blush, which guilty seemed of love.

67

Hope, art thou true, or dost thou flatter me?
Doth Stella now begin with piteous eye
The ruins of her conquest to espy;
Will she take time, before all wracked be?
Her eyes' speech is translated thus by thee:
But fail'st thou not, in phrase so heavenly-high?
Look on again, the fair text better try;
What blushing notes dost thou in margin see?
What sighs stol'n out, or killed before full born?
Hast thou found such, and such-like arguments? 10
Or art thou else to comfort me forsworn?

Well, how so thou interpret the contents,
I am resolved thy error to maintain,
Rather than by more truth to get more pain.

68

Stella, the only planet of my light,
Light of my life, and life of my desire,
Chief good whereto my hope doth only aspire,
World of my wealth, and heaven of my delight;
Why dost thou spend the treasures of thy sprite
With voice more fit to wed Amphion's lyre,
Seeking to quench in me the noble fire
Fed by thy worth, and kindled by thy sight?
And all in vain, for while thy breath most sweet
With choicest words, thy words with reasons rare, 10
Thy reasons firmly set on virtue's feet,
Labour to kill in me this killing care:
O think I then, what paradise of joy
It is, so fair a virtue to enjoy.

69

O joy too high for my low style to show;
O bliss, fit for a nobler state than me;
Envy, put out thine eyes, lest thou do see
What oceans of delight in me do flow.
My friend, that oft saw through all masks my woe,
Come, come, and let me pour myself on thee;
Gone is the winter of my misery,
My spring appears; O see what here doth grow!
For Stella hath, with words where faith doth shine,
Of her high heart giv'n me the monarchy; 10
I, I, O I may say, that she is mine.
And though she give but thus conditionally
This realm of bliss, while virtuous course I take,
No kings be crowned, but they some covenants make.

70

My muse may well grudge at my heavenly joy,
If still I force her in sad rhymes to creep;
She oft hath drunk my tears, now hopes to enjoy

Nectar of mirth, since I Jove's cup do keep.
  Sonnets be not bound prentice to annoy;
Trebles sing high, as well as basses deep;
Grief but love's winter livery is; the boy
Hath cheeks to smile, as well as eyes to weep.
  Come then my muse, show thou height of delight
In well raised notes; my pen the best it may 10
Shall paint out joy, though but in black and white.
Cease, eager muse; peace pen, for my sake stay;
  I give you here my hand for truth of this:
  Wise silence is best music unto bliss.

71

Who will in fairest book of nature know
  How virtue may best lodged in beauty be,
  Let him but learn of love to read in thee,
Stella, those fair lines which true goodness show.
There shall he find all vices' overthrow,
  Not by rude force, but sweetest sovereignty
  Of reason, from whose light those night-birds fly,
That inward sun in thine eyes shineth so.
  And not content to be perfection's heir
Thy self, dost strive all minds that way to move, 10
Who mark in thee what is in thee most fair;
So while thy beauty draws the heart to love,
  As fast thy virtue bends that love to good.
  But ah, desire still cries: 'Give me some food.'

72

Desire, though thou my old companion art,
  And oft so clings to my pure love, that I
  One from the other scarcely can descry,
While each doth blow the fire of my heart;
Now from thy fellowship I needs must part;
  Venus is taught with Dian's wings to fly;
  I must no more in thy sweet passions lie;
Virtue's gold now must head my Cupid's dart.
  Service and honour, wonder with delight,
Fear to offend, will worthy to appear, 10
Care shining in mine eyes, faith in my sprite;

These things are left me by my only dear.
  But thou, desire, because thou would'st have all,
  Now banished art—but yet, alas, how shall?

Second song

Have I caught my heavenly jewel
Teaching sleep most fair to be?
Now will I teach her that she,
When she wakes, is too too cruel.

Since sweet sleep her eyes hath charmed,
The two only darts of love:
Now will I with that boy prove
Some play, while he is disarmed.

Her tongue waking still refuseth,
Giving frankly niggard 'no'; 10
Now will I attempt to know
What 'no' her tongue sleeping useth.

See, the hand which, waking, guardeth,
Sleeping, grants a free resort;
Now will I invade the fort;
Cowards love with loss rewardeth.

But, O fool, think of the danger
Of her just and high disdain;
Now will I, alas, refrain;
Love fears nothing else but anger. 20

Yet those lips so sweetly swelling
Do invite a stealing kiss:
Now will I but venture this;
Who will read, must first learn spelling.

O sweet kiss—but ah, she is waking,
Louring beauty chastens me;
Now will I away hence flee;
Fool, more fool, for no more taking.

73

Love still a boy, and oft a wanton is,
Schooled only by his mother's tender eye;
What wonder then if he his lesson miss,
When for so soft a rod dear play he try?
  And yet my Star, because a sugared kiss
In sport I sucked, while she asleep did lie,
Doth lour, nay chide; nay, threat, for only this.
Sweet, it was saucy love, not humble I.
  But no 'scuse serves, she makes her wrath appear
  In beauty's throne; see now, who dares come near 10
Those scarlet judges, threatening bloody pain?
  O heavenly fool, thy most kiss-worthy face
  Anger invests with such a lovely grace
That anger's self I needs must kiss again.

74

I never drank of Aganippe well,
Nor ever did in shade of Tempe sit;
And muses scorn with vulgar brains to dwell;
Poor layman I, for sacred rites unfit.
  Some do I hear of poet's fury tell,
But (God wot) wot not what they mean by it;
And this I swear, by blackest brook of hell,
I am no pick-purse of another's wit.
  How falls it then, that with so smooth an ease
My thoughts I speak, and what I speak doth flow 10
In verse, and that my verse best wits doth please?
Guess we the cause: 'What, is it thus?' Fie, no;
  'Or so?' Much less. 'How then?' Sure, thus it is:
  My lips are sweet, inspired with Stella's kiss.

75

Of all the kings that ever here did reign,
Edward, named fourth, as first in praise I name;
Not for his fair outside, nor well lined brain,
Although less gifts imp feathers oft on fame;
  Nor that he could, young-wise, wise-valiant, frame
His sire's revenge, joined with a kingdom's gain;
And gained by Mars, could yet mad Mars so tame,

That balance weighed what sword did late obtain;
Nor that he made the flower-de-luce so 'fraid,
Though strongly hedged of bloody lion's paws, 10
That witty Lewis to him a tribute paid;
Nor this, nor that, nor any such small cause;
But only for this worthy knight durst prove
To lose his crown, rather than fail his love.

76

She comes, and straight therewith her shining twins do move
Their rays to me, who in her tedious absence lay
Benighted in cold woe; but now appears my day,
The only light of joy, the only warmth of love.
She comes, with light and warmth, which like Aurora prove
Of gentle force, so that mine eyes dare gladly play
With such a rosy morn, whose beams most freshly gay
Scorch not, but only do dark chilling sprites remove.
But lo, while I do speak, it groweth noon with me;
Her flamy glistering lights increase with time and place; 10
My heart cries, 'Ah, it burns;' mine eyes now dazzled be;
No wind, no shade, can cool; what help then in my case,
But with short breath, long looks, staid feet and walking head,
Pray that my sun go down with meeker beams to bed.

77

Those looks, whose beams be joy, whose motion is delight;
That face, whose lecture shows what perfect beauty is;
That presence, which doth give dark hearts a living light;
That grace, which Venus weeps that she herself doth miss;
That hand, which without touch holds more than Atlas' might;
Those lips, which make death's pay a mean price for a kiss;
That skin, whose pass-praise hue scorns this poor term of 'white';
Those words, which do sublime the quintessence of bliss;
That voice, which makes the soul plant himself in the ears;
That conversation sweet, where such high comforts be, 10
As construed in true speech, the name of heaven it bears,
Makes me in my best thoughts and quiet'st judgement see
That in no more but these I might be fully blessed:
Yet ah, my maiden muse doth blush to tell the rest.

78

O how the pleasant airs of true love be
  Infected by those vapours which arise
  From out that noisome gulf, which gaping lies
Between the jaws of hellish jealousy:
A monster, others' harm, self-misery,
  Beauty's plague, virtue's scourge, succour of lies;
  Who his own joy to his own hurt applies;
And only cherish doth with injury;
  Who, since he hath, by nature's special grace,
  So piercing paws, as spoil when they embrace; 10
So nimble feet, as stir still, though on thorns;
  So many eyes, aye seeking their own woe;
  So ample ears, as never good news know:
Is it not ill that such a devil wants horns?

79

Sweet kiss, thy sweets I fain would sweetly endite,
  Which even of sweetness sweetest sweetener art:
  Pleasing'st consort, where each sense holds a part;
Which, coupling doves, guides Venus' chariot right;
Best charge, and bravest retreat in Cupid's fight;
  A double key, which opens to the heart,
  Most rich, when most his riches it impart;
Nest of young joys, schoolmaster of delight,
  Teaching the mean at once to take and give;
The friendly fray, where blows both wound and heal; 10
The pretty death, while each in other live;
Poor hope's first wealth, hostage of promised weal,
  Breakfast of love—but lo, lo, where she is:
  Cease we to praise, now pray we for a kiss.

80

Sweet swelling lip, well may'st thou swell in pride,
  Since best wits think it wit thee to admire;
  Nature's praise, virtue's stall, Cupid's cold fire,
Whence words, not words, but heavenly graces slide;
The new Parnassus, where the muses bide;
  Sweetener of music, wisdom's beautifier;
  Breather of life, and fastener of desire,

Where beauty's blush in honour's grain is dyed.
  Thus much my heart compelled my mouth to say:
  But now, spite of my heart, my mouth will stay, 10
Loathing all lies, doubting this flattery is,
  And no spur can his resty race renew,
  Without how far this praise is short of you,
Sweet lip, you teach my mouth with one sweet kiss.

81

O kiss, which dost those ruddy gems impart,
Or gems, or fruits of new-found paradise,
Breathing all bliss, and sweetening to the heart,
Teaching dumb lips a nobler exercise;
  O kiss, which souls, even souls together ties
By links of love, and only nature's art;
How fain would I paint thee to all men's eyes,
Or of thy gifts at least shade out some part.
  But she forbids; with blushing words, she says
  She builds her fame on higher seated praise; 10
But my heart burns, I cannot silent be.
  Then since (dear life) you fain would have me peace,
  And I, mad with delight, want wit to cease,
Stop you my mouth with still still kissing me.

82

Nymph of the garden where all beauties be;
  Beauties, which do in excellency pass
  His who till death looked in a watery glass,
Or hers whom naked the Trojan boy did see:
Sweet garden nymph, which keeps the cherry tree,
  Whose fruit doth far th'Hesperian taste surpass;
  Most sweet-fair, most fair-sweet, do not, alas,
From coming near those cherries banish me.
  For though, full of desire, empty of wit,
Admitted late by your best-graced grace, 10
I caught at one of them a hungry bit;
Pardon that fault, once more grant me the place,
  And I do swear, even by the same delight,
  I will but kiss, I never more will bite.

83

Good brother Philip, I have borne you long;
  I was content you should in favour creep,
  While craftily you seemed your cut to keep,
As though that fair soft hand did you great wrong.
I bare (with envy) yet I bare your song,
  When in her neck you did love-ditties peep;
  Nay, more fool I, oft suffered you to sleep
In lilies' nest, where love's self lies along.
  What, doth high place ambitious thoughts augment?
Is sauciness reward of courtesy? 10
Cannot such grace your silly self content,
But you must needs with those lips billing be,
  And through those lips drink nectar from that tongue?
  Leave that, sir Phip, lest off your neck be wrung.

Third song

If Orpheus' voice had force to breathe such music's love
Through pores of senseless trees, as it could make them move;
If stones good measure danced, the Theban walls to build,
To cadence of the tunes, which Amphion's lyre did yield;
More cause a like effect at leastwise bringeth:
O stones, O trees, learn hearing: Stella singeth.

If love might sweeten so a boy of shepherd brood,
To make a lizard dull to taste love's dainty food;
If eagle fierce could so in Grecian maid delight,
As his light was her eyes, her death his endless night; 10
Earth gave that love, Heaven I trow love refineth:
O birds, O beasts, look, love: lo, Stella shineth.

The birds, beasts, stones, and trees feel this, and feeling, love;
And if the trees, nor stones, stir not, the same to prove,
Nor beasts nor birds do come unto this blessed gaze,
Know, that small love is quick, and great love doth amaze:
They are amazed, but you with reason armed:
O eyes, O ears of men, how are you charmed!

84

Highway, since you my chief Parnassus be,
And that my muse, to some ears not unsweet,
Tempers her words to trampling horse's feet
More oft than to a chamber melody;
Now blessed you, bear onward blessed me
To her, where I my heart safeliest shall meet.
My muse and I must you of duty greet,
With thanks and wishes, wishing thankfully.
Be you still fair, honoured by public heed,
By no encroachment wronged, nor time forgot; 10
Nor blamed for blood, nor shamed for sinful deed.
And that you know, I envy you no lot
Of highest wish, I wish you so much bliss,
Hundreds of years you Stella's feet may kiss.

85

I see the house; my heart, thy self contain;
Beware full sails drown not thy tottering barge,
Lest joy, by nature apt sprites to enlarge,
Thee to thy wrack beyond thy limits strain;
Nor do like lords, whose weak confused brain,
Not pointing to fit folks each undercharge,
While every office themselves will discharge,
With doing all, leave nothing done but pain.
But give apt servants their due place; let eyes
See beauty's total sum summed in her face; 10
Let ears hear speech, which wit to wonder ties;
Let breath suck up those sweets; let arms embrace
The globe of weal; lips love's indentures make;
Thou but of all the kingly tribute take.

Fourth song

Only joy, now here you are,
Fit to hear and ease my care;
Let my whispering voice obtain
Sweet reward for sharpest pain:
Take me to thee and thee to me.
'No, no, no, no, my dear, let be.'

Night hath closed all in her cloak,
Twinkling stars love-thoughts provoke;
Danger hence good care doth keep;
Jealousy itself doth sleep: 10
Take me to thee and thee to me.
'No, no, no, no, my dear, let be.'

Better place no wit can find
Cupid's yoke to loose or bind;
These sweet flowers on fine bed too
Us in their best language woo:
Take me to thee and thee to me.
'No, no, no, no, my dear, let be.'

This small light the moon bestows
Serves thy beams but to disclose, 20
So to raise my hap more high;
Fear not else, none can us spy:
Take me to thee and thee to me.
'No, no, no, no, my dear, let be.'

That you heard was but a mouse;
Dumb sleep holdeth all the house;
Yet asleep, methinks, they say,
Young folks, take time while you may:
Take me to thee and thee to me.
'No, no, no, no, my dear, let be.' 30

Niggard time threats, if we miss
This large offer of our bliss
Long stay ere he grant the same;
Sweet then, while each thing doth frame:
Take me to thee and thee to me.
'No, no, no, no, my dear, let be.'

Your fair mother is abed,
Candles out, and curtains spread;
She thinks you do letters write;
Write, but first let me endite:
Take me to thee and thee to me.
'No, no, no, no, my dear, let be.'

Sweet, alas, why strive you thus?
Concord better fitteth us.
Leave to Mars the force of hands,
Your power in your beauty stands:
Take me to thee and thee to me.
'No, no, no, no, my dear, let be.'

Woe to me, and do you swear
Me to hate, but I forbear? 50
Cursed be my destinies all,
That brought me so high, to fall;
Soon with my death I will please thee.
'No, no, no, no, my dear, let be.'

86

Alas, whence came this change of looks? If I
Have changed desert, let mine own conscience be
A still felt plague, to self condemning me:
Let woe gripe on my heart, shame load mine eye.
But if all faith, like spotless ermine, lie
Safe in my soul, which only doth to thee
(As his sole object to felicity)
With wings of love in air of wonder fly,
O ease your hand, treat not so hard your slave;
In justice pains come not till faults do call; 10
Or if I needs, sweet judge, must torments have,
Use something else to chasten me withal
Than those blessed eyes, where all my hopes do dwell.
No doom should make one's heaven become his hell.

Fifth song

While favour fed my hope, delight with hope was brought;
Thought waited on delight, and speech did follow thought;
Then grew my tongue and pen records unto thy glory;
I thought all words were lost, that were not spent of thee;
I thought each place was dark but where thy lights would be,
And all ears worse than deaf, that heard not out thy story.

I said thou wert most fair, and so indeed thou art;
I said thou wert most sweet, sweet poison to my heart;
I said my soul was thine—O that I then had lied!
I said thine eyes were stars, thy breasts the milken way, 10
Thy fingers Cupid's shafts, thy voice the angels' lay,
And all I said so well, as no man it denied.

But now that hope is lost, unkindness kills delight,
Yet thought and speech do live, though metamorphosed quite;
For rage now rules the reins, which guided were by pleasure.
I think now of thy faults, who late thought of thy praise;
That speech falls now to blame, which did thy honour raise;
The same key open can, which can lock up a treasure.

Thou then, whom partial heavens conspired in one to frame,
The proof of beauty's worth, th'inheritrix of fame, 20
The mansion seat of bliss, and just excuse of lovers;
See now those feathers plucked, wherewith thou flew'st most high;
See what clouds of reproach shall dark thy honour's sky;
Whose own fault casts him down, hardly high seat recovers.

And O my muse, though oft you lulled her in your lap,
And then, a heavenly child, gave her ambrosian pap,
And to that brain of hers your hidd'nest gifts infused;
Since she, disdaining me, doth you in me disdain,
Suffer her not to laugh, while we both suffer pain;
Princes in subjects wronged, must deem themselves abused. 30

Your client poor my self, shall Stella handle so?
Revenge, revenge, my muse; defiance' trumpet blow;
Threaten what may be done, yet do more than you threaten.
Ah, my suit granted is; I feel my breast to swell;
Now child, a lesson new you shall begin to spell:
Sweet babes must babies have, but shrewd girls must be beaten.

Think now no more to hear of warm fine-odoured snow,
Nor blushing lilies, nor pearls' ruby-hidden row,
Nor of that golden sea, whose waves in curls are broken:
But of thy soul, so fraught with such ungratefulness, 40
As where thou soon might'st help, most faith doth most oppress;
Ungrateful who is called, the worst of evils is spoken.

Yet worse than worst, I say thou art a thief. A thief?
No God forbid. A thief, and of worst thieves the chief;
Thieves steal for need, and steal but goods, which pain recovers,
But thou, rich in all joys, dost rob my joys from me,
Which cannot be restored by time nor industry.
Of foes the spoil is evil, far worse of constant lovers.

Yet gentle English thieves do rob, but will not slay;
Thou, English murdering thief, wilt have hearts for thy prey; 50
The name of 'murderer' now on thy fair forehead sitteth;
And even while I do speak, my death wounds bleeding be,
Which, I protest, proceed from only cruel thee.
Who may, and will not, save, murder in truth committeth.

But murder, private fault, seems but a toy to thee;
I lay then to thy charge, unjustest tyranny,
If rule by force without all claim a tyrant showeth.
For thou dost lord my heart, who am not born thy slave;
And which is worse, makes me, most guiltless, torments have;
A rightful prince by unright deeds a tyrant groweth. 60

Lo, you grow proud with this, for tyrants make folk bow.
Of foul rebellion then I do appeach thee now;
Rebel by nature's law, rebel by law of reason.
Thou, sweetest subject, wert born in the realm of love,
And yet against thy prince thy force dost daily prove;
No virtue merits praise, once touched with blot of treason.

But valiant rebels oft in fools' mouths purchase fame;
I now then stain thy white with vagabonding shame,
Both rebel to the son, and vagrant from the mother:
For wearing Venus' badge in every part of thee 70
Unto Diana's train thou, runaway, did'st flee:
Who faileth one, is false, though trusty to another.

What, is not this enough? Nay, far worse cometh here:
A witch I say thou art, though thou so fair appear;
For I protest, my sight never thy face enjoyeth,
But I in me am changed; I am alive and dead;
My feet are turned to roots; my heart becometh lead.
No witchcraft is so evil, as which man's mind destroyeth.

Yet witches may repent; thou art far worse than they;
Alas, that I am forced such evil of thee to say! 80
I say thou art a devil, though clothed in angel's shining;
For thy face tempts my soul to leave the heaven for thee,
And thy words of refuse do pour even hell on me:
Who tempt, and tempted plague, are devils in true defining.

You then, ungrateful thief; you murdering tyrant, you;
You rebel runaway, to lord and lady untrue;
You witch, you devil, alas—you still of me beloved;
You see what I can say; mend yet your froward mind,
And such skill in my muse you, reconciled, shall find,
That all these cruel words your praises shall be proved. 90

Sixth song

O you that hear this voice,
O you that see this face,
Say whether of the choice
Deserves the former place:
Fear not to judge this bate,
For it is void of hate.

This side doth beauty take;
For that, doth music speak;
Fit orators to make
The strongest judgements weak: 10
The bar to plead their right
Is only true delight.

Thus doth the voice and face,
These gentle lawyers, wage,
Like loving brothers' case
For father's heritage,
That each, while each contends,
Itself to other lends.

For beauty beautifies
With heavenly hue and grace 20
The heavenly harmonies;

And in this faultless face
The perfect beauties be
A perfect harmony.

Music more lofty swells
In speeches nobly placed;
Beauty as far excels
In action aptly graced;
A friend each party draws
To countenance his cause. 30

Love more affected seems
To beauty's lovely light,
And wonder more esteems
Of music's wondrous might;
But both to both so bent
As both in both are spent.

Music doth witness call
The ear, his truth to try;
Beauty brings to the hall
The judgement of the eye: 40
Both in their objects such,
As no exceptions touch.

The common sense, which might
Be arbiter of this,
To be, forsooth, upright,
To both sides partial is:
He lays on this chief praise,
Chief praise on that he lays.

Then reason, princess high,
Whose throne is in the mind, 50
Which music can in sky
And hidden beauties find:
Say whether thou wilt crown
With limitless renown.

Seventh song

Whose senses in so ill consort their stepdame nature lays,
That ravishing delight in them most sweet tunes do not raise;
Or if they do delight therein, yet are so cloyed with wit,
As with sententious lips to set a title vain on it;
O let them hear these sacred tunes, and learn in wonder's
  schools
To be, in things past bounds of wit, fools, if they be not fools.

Who have so leaden eyes, as not to see sweet beauty's show;
Or seeing, have so wooden wits, as not that worth to know;
Or knowing, have so muddy minds, as not to be in love;
Or loving, have so frothy thoughts as eas'ly thence to move: 10
O, let them see these heavenly beams, and in fair letters read
A lesson fit, both sight and skill, love and firm love to breed.

Hear then, but then with wonder hear; see, but adoring see;
No mortal gifts, no earthly fruits, now here descended be;
See; do you see this face? A face? Nay, image of the skies,
Of which the two life-giving lights are figured in her eyes.
Hear you this soul-invading voice, and count it but a voice?
The very essence of their tunes, when angels do rejoice.

Eighth song

In a grove most rich of shade,
Where birds wanton music made,
May then young his pied weeds showing,
New perfumed with flowers fresh growing,

Astrophil with Stella sweet
Did for mutual comfort meet;
Both within themselves oppressed,
But each in the other blessed.

Him great harms had taught much care:
Her fair neck a foul yoke bare: 10
But her sight his cares did banish,
In his sight her yoke did vanish.

Wept they had, alas the while;
But now tears themselves did smile,
While their eyes, by love directed,
Interchangeably reflected.

Sigh they did; but now betwixt
Sighs of woes were glad sighs mixed,
With arms crossed, yet testifying
Restless rest, and living dying. 20

Their ears hungry of each word,
Which the dear tongue would afford,
But their tongues restrained from walking,
Till their hearts had ended talking.

But when their tongues could not speak
Love itself did silence break;
Love did set his lips asunder,
Thus to speak in love and wonder:

'Stella, sovereign of my joy,
Fair triumpher of annoy, 30
Stella, star of heavenly fire,
Stella, lodestar of desire;

'Stella, in whose shining eyes
Are the lights of Cupid's skies;
Whose beams, where they once are darted,
Love therewith is straight imparted;

'Stella, whose voice when it speaks
Senses all asunder breaks;
Stella, whose voice when it singeth
Angels to acquaintance bringeth; 40

'Stella, in whose body is
Writ each character of bliss;
Whose face all, all beauty passeth,
Save thy mind, which yet surpasseth:

'Grant, O grant—but speech, alas,
Fails me, fearing on to pass;
Grant—O me, what am I saying?
But no fault there is in praying:

'Grant, O dear, on knees I pray'—
(Knees on ground he then did stay) 50
'That not I, but since I love you,
Time and place for me may move you.

'Never season was more fit,
Never room more apt for it;
Smiling air allows my reason;
These birds sing, "Now use the season";

'This small wind, which so sweet is,
See how it the leaves doth kiss,
Each tree in his best attiring
Sense of love to love inspiring. 60

'Love makes earth the water drink,
Love to earth makes water sink;
And if dumb things be so witty,
Shall a heavenly grace want pity?'

There his hands in their speech fain
Would have made tongue's language plain:
But her hands, his hands repelling,
Gave repulse, all grace excelling.

Then she spake; her speech was such
As not ears, but heart did touch; 70
While such wise she love denied,
As yet love she signified.

'Astrophil,' said she, 'my love
'Cease in these effects to prove:
Now be still; yet still believe me,
Thy grief more than death would grieve me.

'If that any thought in me
Can taste comfort but of thee,
Let me, fed with hellish anguish,
Joyless, hopeless, endless languish. 80

'If those eyes you praised be
Half so dear as you to me,
Let me home return, stark blinded
Of those eyes, and blinder minded.

'If to secret of my heart
I do any wish impart
Where thou art not foremost placed,
Be both wish and I defaced.

'If more may be said, I say:
All my bliss in thee I lay; 90
If thou love, my love content thee,
For all love, all faith is meant thee.

'Trust me, while I thee deny,
In my self the smart I try;
Tyrant honour thus doth use thee;
Stella's self might not refuse thee.

'Therefore, dear, this no more move,
Lest, though I leave not thy love,
Which too deep in me is framed,
I should blush when thou art named.' 100

Therewithal away she went,
Leaving him so passion-rent
With what she had done and spoken,
That therewith my song is broken.

Ninth song

Go, my flock, go get you hence,
Seek a better place of feeding,
Where you may have some defence

From the storms in my breast breeding,
And showers from my eyes proceeding.

Leave a wretch, in whom all woe
Can abide to keep no measure;
Merry flock, such one forego,
Unto whom mirth is displeasure,
Only rich in mischief's treasure. 10

Yet, alas, before you go,
Hear your woeful master's story,
Which to stones I else would show:
Sorrow only then hath glory,
When 'tis excellently sorry.

Stella, fiercest shepherdess,
Fiercest, but yet fairest ever;
Stella, whom, O heavens, do bless,
Though against me she persevere,
Though I bliss inherit never; 20

Stella hath refused me,
Stella, who more love hath proved
In this caitiff heart to be
Than can in good ewes be moved
Toward lambkins best beloved.

Stella hath refused me;
Astrophil, that so well served,
In this pleasant spring must see,
While in pride flowers be preserved,
Himself only winter-starved. 30

Why, alas, doth she then swear
That she loveth me so dearly,
Seeing me so long to bear
Coals of love, that burn so clearly,
And yet leave me helpless merely?

Is that love? Forsooth, I trow,
If I saw my good dog grieved,

And a help for him did know, 40
My love should not be believed
But he were by me relieved.

No, she hates me, wellaway,
Faining love somewhat, to please me;
For she knows, if she display
All her hate, death soon would seize me,
And of hideous torments ease me.

Then adieu, dear flock, adieu:
But alas, if in your straying
Heavenly Stella meet with you,
Tell her, in your piteous blaying,
Her poor slave's unjust decaying. 50

87

When I was forced from Stella, ever dear,
Stella, food of my thoughts, heart of my heart,
Stella, whose eyes make all my tempests clear,
By iron laws of duty to depart;
  Alas, I found that she with me did smart,
I saw that tears did in her eyes appear;
I saw that sighs her sweetest lips did part,
And her sad words my sadded sense did hear.
  For me, I wept, to see pearls scattered so;
  I sighed her sighs, and wailed for her woe; 10
Yet swam in joy, such love in her was seen.
  Thus while th'effect most bitter was to me,
  And nothing than the cause more sweet could be,
I had been vexed, if vexed I had not been.

88

Out, traitor absence; darest thou counsel me
From my dear captainess to run away,
Because in brave array here marcheth she
That to win me, oft shows a present pay?
  Is faith so weak? Or is such force in thee?
When sun is hid, can stars such beams display?
Cannot heaven's food, once felt, keep stomachs free

From base desire on earthly cates to prey?
  Tush, absence; while thy mists eclipse that light,
  My orphan sense flies to the inward sight, 10
Where memory sets forth the beams of love;
  That where before heart loved and eyes did see,
  In heart both sight and love now coupled be;
United powers make each the stronger be.

89

Now that of absence the most irksome night
  With darkest shade doth overcome my day;
  Since Stella's eyes, wont to give me my day,
Leaving my hemisphere, leave me in night;
Each day seems long, and longs for long-stayed night;
  The night as tedious, woos the approach of day;
  Tired with the dusty toils of busy day,
Languished with horrors of the silent night,
Suffering the ills both of the day and night,
  While no night is more dark than is my day, 10
Nor no day hath less quiet than my night;
  With such bad mixture of my night and day,
That living thus in blackest winter night,
  I feel the flames of hottest summer day.

90

Stella, think not that I by verse seek fame;
  Who seek, who hope, who love, who live, but thee:
  Thine eyes my pride, thy lips my history;
If thou praise not, all other praise is shame.
Nor so ambitious am I, as to frame
  A nest for my young praise in laurel tree;
  In truth I swear, I wish not there should be
Graved in mine epitaph a poet's name:
  Ne if I would, could I just title make,
That any laud to me thereof should grow, 10
Without my plumes from others' wings I take.
For nothing from my wit or will doth flow,
  Since all my words thy beauty doth endite,
  And love doth hold my hand, and makes me write.

91

Stella, while now, by honour's cruel might,
  I am from you, light of my life, misled,
  And that fair you, my sun, thus overspread
With absence' veil, I live in sorrow's night;
If this dark place yet show, like candle light,
  Some beauty's piece, as amber-coloured head,
  Milk hands, rose cheeks, or lips more sweet, more red,
Or seeing jets, black, but in blackness bright:
  They please, I do confess, they please mine eyes.
But why? Because of you they models be, 10
Models such be wood-globes of glistering skies.
Dear, therefore be not jealous over me;
  If you hear that they seem my heart to move,
  Not them, O no, but you in them I love.

92

Be your words made, good sir, of Indian ware,
  That you allow me them by so small rate?
  Or do you cutted Spartans imitate?
Or do you mean my tender ears to spare,
That to my questions you so total are?
  When I demand of Phoenix Stella's state,
  You say, forsooth, you left her well of late.
O God, think you that satisfies my care?
  I would know whether she did sit or walk,
How clothed, how waited on? Sighed she or smiled? 10
Whereof, with whom, how often did she talk?
With what pastime time's journey she beguiled?
  If her lips deigned to sweeten my poor name?
  Say all, and all well said, still say the same.

Tenth song

O dear life, when shall it be
  That mine eyes thine eyes may see,
  And in them my mind discover,
  Whether absence have had force
  Thy remembrance to divorce
  From the image of thy lover?

O if I myself find not
  After parting aught forgot,
  Nor debarred from beauty's treasure,
  Let no tongue aspire to tell 10
  In what high joys I shall dwell;
  Only thought aims at the pleasure.

Thought, therefore, I will send thee
  To take up the place for me
  Long I will not after tarry.
  There unseen thou may'st be bold
  Those fair wonders to behold,
  Which in them my hopes do carry.

Thought, see thou no place forbear;
  Enter bravely everywhere, 20
  Seize on all to her belonging;
  But if thou would'st guarded be,
  Fearing her beams, take with thee
  Strength of liking, rage of longing.

Think of that most grateful time
  When my leaping heart will climb
  In my lips to have his biding,
  There those roses for to kiss
  Which do breathe a sugared bliss, 30
  Opening rubies, pearls dividing.

Think of my most princely power,
  When I, blessed, shall devour
  With my greedy lickerous senses,
  Beauty, music, sweetness, love,
  While she doth against me prove
  Her strong darts but weak defences.

Think, think of those dallyings,
  When with dove-like murmurings,
  With glad moaning passed anguish, 40
  We change eyes, and heart for heart
  Each to other do impart,
  Joying, till joy make us languish.

O my thought, my thoughts surcease;
Thy delights my woes increase,
My life melts with too much thinking.
Think no more, but die in me,
Till thou shalt revived be
At her lips my nectar drinking.

93

O fate, O fault, O curse, child of my bliss;
What sobs can give words grace my grief to show?
What ink is black enough to paint my woe?
Through me, wretch me, even Stella vexed is.
Yet truth—if caitiff's breath might call thee—this
Witness with me, that my foul stumbling so
From carelessness did in no manner grow;
But wit, confused with too much care, did miss.
And do I then myself this vain 'scuse give?
I have (live I, and know this?) harmed thee; 10
Though worlds 'quit me, shall I myself forgive?
Only with pains my pains thus eased be,
That all my hurts in my heart's wrack I read;
I cry thy sighs; my dear, thy tears I bleed.

94

Grief, find the words; for thou hast made my brain
So dark with misty vapours, which arise
From out thy heavy mould, that inbent eyes
Can scarce discern the shape of mine own pain.
Do thou then (for thou canst), do thou complain,
For my poor soul, which now that sickness tries
Which even to sense, sense of itself denies,
Though harbingers of death lodge there his train.
Or if thy love of plaint yet mine forbears,
As of a caitiff, worthy so to die; 10
Yet wail thyself, and wail with causeful tears,
That though in wretchedness thy life doth lie,
Yet grow'st more wretched than thy nature bears,
By being placed in such a wretch as I.

95

Yet sighs, dear sighs, indeed true friends you are,
  That do not leave your least friend at the worst;
  But as you with my breast I oft have nursed,
So grateful now you wait upon my care.
Faint coward joy no longer tarry dare,
  Seeing hope yield when this woe strake him first;
  Delight protests he is not for the accursed,
Though oft himself my mate-in-arms he sware.
  Nay, sorrow comes with such main rage, that he
Kills his own children, tears, finding that they 10
By love were made apt to consort with me.
Only true sighs, you do not go away:
  Thank may you have for such a thankful part,
  Thank-worthiest yet, when you shall break my heart.

96

Thought, with good cause thou lik'st so well the night,
  Since kind or chance gives both one livery:
  Both sadly black, both blackly darkened be,
Night barred from sun, thou from thy own sun's light.
Silence in both displays his sullen might;
  Slow heaviness in both holds one degree,
  That full of doubts, thou of perplexity;
Thy tears express night's native moisture right.
  In both a mazeful solitariness:
In night, of sprites the ghastly powers stir, 10
In thee, or sprites, or sprited ghastliness,
But, but, alas, night's side the odds hath, far,
  For that at length yet doth invite some rest,
  Thou, though still tired, yet still dost it detest.

97

Dian, that fain would cheer her friend, the night,
  Shows her oft at the full her fairest face,
  Bringing with her those starry nymphs, whose chase
From heavenly standing hits each mortal wight.
But ah, poor night, in love with Phoebus' light,
  And endlessly despairing of his grace,
  Herself (to show no other joy hath place)

Silent and sad, in mourning weeds doth dight:
  Even so, alas, a lady, Dian's peer,
With choice delights and rarest company 10
Would fain drive clouds from out my heavy cheer.
But woe is me, though joy itself were she,
  She could not show my blind brain ways of joy,
  While I despair my sun's sight to enjoy.

98

Ah bed, the field where joy's peace some do see,
  The field where all my thoughts to war be trained,
  How is thy grace by my strange fortune stained!
How thy lee shores by my sighs stormed be!
With sweet soft shades thou oft invitest me
  To steal some rest; but, wretch, I am constrained
  (Spurred with love's spur, though galled and shortly reined
With care's hard hand) to turn and toss in thee,
  While the black horrors of the silent night
  Paint woe's black face so lively to my sight 10
That tedious leisure marks each wrinkled line.
  But when Aurora leads out Phoebus' dance,
  Mine eyes then only wink, for spite, perchance,
That worms should have their sun, and I want mine.

99

When far spent night persuades each mortal eye,
  To whom nor art nor nature granteth light,
  To lay his then mark-wanting shafts of sight,
Closed with their quivers, in sleep's armoury;
With windows ope then most my mind doth lie,
  Viewing the shape of darkness and delight,
  Takes in that sad hue, which with the inward night
Of his mazed powers keeps perfect harmony.
  But when birds charm, and that sweet air, which is
Morn's messenger, with rose-enamelled skies, 10
Calls each wight to salute the flower of bliss:
In tomb of lids then buried are mine eyes,
  Forced by their lord, who is ashamed to find
  Such light in sense, with such a darkened mind.

100

O tears, no tears, but rain from beauty's skies,
Making those lilies and those roses grow
Which aye most fair, now more than most fair show,
While graceful pity beauty beautifies:
O honeyed sighs, which from that breast do rise
Whose pants do make unspilling cream to flow,
Winged with whose breath so pleasing zephyrs blow,
As can refresh the hell where my soul fries:
O plaints, conserved in such a sugared phrase
That eloquence itself envies your praise, 10
While sobbed-out words a perfect music give:
Such tears, sighs, plaints, no sorrow is, but joy;
Or if such heavenly signs must prove annoy,
All mirth farewell, let me in sorrow live.

101

Stella is sick, and in that sick-bed lies
Sweetness, that breathes and pants as oft as she;
And grace, sick too, such fine conclusions tries
That sickness brags itself best graced to be.
Beauty is sick, but sick in so fair guise
That in that paleness beauty's white we see;
And joy, which is inseparate from these eyes,
Stella now learns (strange case!) to weep in thee.
Love moves thy pain, and like a faithful page,
As thy looks stir, runs up and down, to make 10
All folks prest at thy will thy pain to assuage;
Nature with care sweats for her darling's sake,
Knowing worlds pass, ere she enough can find
Of such heaven stuff, to clothe so heavenly mind.

102

Where be those roses gone, which sweetened so our eyes?
Where those red cheeks, which oft with fair increase did frame
The height of honour in the kindly badge of shame?
Who hath the crimson weeds stol'n from my morning skies?
How doth the colour vade of those vermilion dyes,
Which nature's self did make, and self engrained the same?
I would know by what right this paleness overcame

That hue, whose force my heart still unto thraldom ties?
Galen's adoptive sons, who by a beaten way
Their judgements hackney on, the fault on sickness lay; 10
But feeling proof makes me say they mistake it far:
It is but love, which makes his paper perfect white
To write therein more fresh the story of delight,
While beauty's reddest ink Venus for him doth stir.

103

O happy Thames, that didst my Stella bear!
I saw thyself, with many a smiling line
Upon thy cheerful face, joy's livery wear,
While those fair planets on thy streams did shine.
The boat for joy could not to dance forebear,
While wanton winds, with beauties so divine
Ravished, stayed not, till in her golden hair
They did themselves (O sweetest prison!) twine.
And fain those Aeol's youths there would their stay
Have made; but forced by nature still to fly, 10
First did with puffing kiss those locks display.
She, so dishevelled, blushed; from window I
With sight thereof cried out, 'O fair disgrace;
Let honour's self to thee grant highest place.'

104

Envious wits, what hath been mine offence,
That with such poisonous care my looks you mark,
That to each word, nay, sigh, of mine you hark,
As grudging me my sorrow's eloquence?
Ah, is it not enough, that I am thence,
Thence, so far thence, that scarcely any spark
Of comfort dare come to this dungeon dark,
Where rigorous exile locks up all my sense?
But if I by a happy window pass;
If I but stars upon my armour bear; 10
Sick, thirsty, glad, though but of empty glass;
Your moral notes straight my hid meaning tear
From out my ribs, and puffing prove that I
Do Stella love. Fools, who doth it deny?

Eleventh song

'Who is it that this dark night
Underneath my window plaineth?'
It is one that from thy sight
Being, ah, exiled, disdaineth
Every other vulgar light.

'Why, alas, and are you he?
Be not yet those fancies changed?'
Dear, when you find change in me,
Though from me you be estranged,
Let my change to ruin be. 10

'Well, in absence this will die;
Leave to see, and leave to wonder.'
Absence sure will help, if I
Can learn, how myself to sunder
From what in my heart doth lie.

'But time will these thoughts remove;
Time doth work what no man knoweth.'
Time doth as the subject prove;
With time still the affection groweth
In the faithful turtle dove. 20

'What if you new beauties see,
Will not they stir new affection?'
I will think they pictures be,
Image-like of saints' perfection,
Poorly counterfeiting thee.

'But your reason's purest light
Bids you leave such minds to nourish.'
Dear, do reason no such spite;
Never doth thy beauty flourish
More than in my reason's sight. 30

'But the wrongs love bears, will make
Love at length leave undertaking.'

No, the more fools do it shake
In a ground of so firm making
Deeper still they drive the stake.

'Peace, I think that some give ear;
Come no more, lest I get anger.'
Bliss, I will my bliss forebear,
Fearing, sweet, you to endanger,
But my soul shall harbour there. 40

'Well, be gone, be gone, I say,
Lest that Argus' eyes perceive you.'
O, unjust is fortune's sway,
Which can make me thus to leave you,
And from louts to run away.

105

Unhappy sight, and hath she vanished by,
So near, in so good time, so free a place?
Dead glass, dost thou thy object so embrace
As what my heart still sees, thou canst not spy?
I swear by her I love and lack, that I
Was not in fault, who bent the dazzling race
Only unto the heaven of Stella's face,
Counting but dust what in the way did lie.
But cease, mine eyes, your tears do witness well
That you, guiltless thereof, your nectar missed. 10
Cursed be the page from whom the bad torch fell,
Cursed be the night which did your strife resist,
Cursed be the coachman, which did drive so fast,
With no worse curse than absence makes me taste.

106

O absent presence, Stella is not here;
False flattering hope, that with so fair a face
Bare me in hand, that in this orphan place
Stella, I say my Stella, should appear.
What say'st thou now? Where is that dainty cheer
Thou told'st mine eyes should help their famished case?
But thou art gone, now that self-felt disgrace

Doth make me most to wish thy comfort near.
But here I do store of fair ladies meet,
Who may with charm of conversation sweet 10
Make in my heavy mould new thoughts to grow:
Sure they prevail as much with me, as he
That bade his friend, but then new maimed, to be
Merry with him, and not think of his woe.

107

Stella, since thou so right a princess art
Of all the powers which life bestows on me,
That ere by them aught undertaken be
They first resort unto that sovereign part;
Sweet, for a while give respite to my heart,
Which pants as though it still should leap to thee;
And on my thoughts give thy lieutenancy
To this great cause, which needs both use and art;
And as a queen, who from her presence sends
Whom she employs, dismiss from thee my wit, 10
Till it have wrought what thy own will attends.
On servants' shame oft master's blame doth sit;
O, let not fools in me thy works reprove,
And scorning say, 'See, what it is to love!'

108

When sorrow, using mine own fire's might,
Melts down his lead into my boiling breast,
Through that dark furnace to my heart oppressed
There shines a joy from thee, my only light;
But soon as thought of thee breeds my delight,
And my young soul flutters to thee, his nest;
Most rude despair, my daily unbidden guest,
Clips straight my wings, straight wraps me in his night,
And makes me then bow down my head, and say:
'Ah, what doth Phoebus' gold that wretch avail 10
Whom iron doors do keep from use of day?'
So strangely, alas, thy works in me prevail,
That in my woes for thee thou art my joy,
And in my joys for thee my only annoy.

THE DEFENCE OF POESY

WHEN the right virtuous Edward Wotton and I were at the Emperor's court together, we gave ourselves to learn horsemanship of John Pietro Pugliano, one that with great commendation had the place of an esquire in his stable. And he, according to the fertileness of the Italian wit, did not only afford us the demonstration of his practice, but sought to enrich our minds with the contemplations therein, which he thought most precious. But with none I remember mine ears were at that time more laden, than when (either angered with slow payment, or moved with our learner-like admira-
10 tion) he exercised his speech in the praise of his faculty. He said soldiers were the noblest estate of mankind, and horsemen the noblest of soldiers. He said they were the masters of war and ornaments of peace, speedy goers and strong abiders, triumphers both in camps and courts. Nay, to so unbelieved a point he proceeded as that no earthly thing bred such wonder to a prince as to be a good horseman—skill of government was but a *pedanteria* in comparison. Then would he add certain praises, by telling what a peerless beast the horse was, the only serviceable courtier without flattery, the beast of most beauty, faithfulness, courage, and such
20 more, that if I had not been a piece of a logician before I came to him, I think he would have persuaded me to have wished myself a horse. But thus much at least with his no few words he drave into me, that self-love is better than any gilding to make that seem gorgeous wherein ourselves be parties. Wherein, if Pugliano's strong affection and weak arguments will not satisfy you, I will give you a nearer example of myself, who (I know not by what mischance) in these my not old years and idlest times having slipped into the title of a poet, am provoked to say something unto you in the defence of that my unelected vocation, which if I handle with
30 more good will than good reasons, bear with me, since the scholar is to be pardoned that followeth the steps of his master. And yet I must say that, as I have more just cause to make a pitiful defence of poor poetry, which from almost the highest estimation of learning is fallen to be the laughing-stock of children, so have I need to bring some more available proofs: since the former is by no man barred of his deserved credit, the silly latter hath had even the

names of philosophers used to the defacing of it, with great danger of civil war among the Muses.

And first, truly, to all them that, professing learning, inveigh against poetry, may justly be objected that they go very near to 40 ungratefulness, to seek to deface that which, in the noblest nations and languages that are known, hath been the first light-giver to ignorance, and first nurse, whose milk by little and little enabled them to feed afterwards of tougher knowledges. And will they now play the hedgehog that, being received into the den, drave out his host? Or rather the vipers, that with their birth kill their parents?

Let learned Greece in any of his manifold sciences be able to show me one book before Musaeus, Homer, and Hesiod, all three nothing else but poets. Nay, let any history be brought that can say any writers were there before them, if they were not men of the 50 same skill, as Orpheus, Linus, and some other are named, who, having been the first of that country that made pens deliverers of their knowledge to the posterity, may justly challenge to be called their fathers in learning: for not only in time they had this priority (although in itself antiquity be venerable) but went before them, as causes to draw with their charming sweetness the wild untamed wits to an admiration of knowledge. So, as Amphion was said to move stones with his poetry to build Thebes, and Orpheus to be listened to by beasts, indeed, stony and beastly people; so among the Romans were Livius Andronicus and Ennius. So in the Italian 60 language the first that made it aspire to be a treasure-house of science were the poets Dante, Boccaccio, and Petrarch. So in our English were Gower and Chaucer, after whom, encouraged and delighted with their excellent fore-going, others have followed, to beautify our mother tongue, as well in the same kind as in other arts.

This did so notably show itself, that the philosophers of Greece durst not a long time appear to the world but under the masks of poets. So Thales, Empedocles, and Parmenides sang their natural philosophy in verses; so did Pythagoras and Phocylides their moral 70 counsels; so did Tyrtaeus in war matters, and Solon in matters of policy: or rather they, being poets, did exercise their delightful vein in those points of highest knowledge, which before them lay hid to the world. For that wise Solon was directly a poet it is manifest, having written in verse the notable fable of the Atlantic Island, which was continued by Plato. And truly even Plato whosoever well considereth shall find that in the body of his work, though the inside and strength were philosophy, the skin, as it were, and beauty depended most of poetry: for all standeth upon dialogues, wherein 80 he feigneth many honest burgesses of Athens to speak of such matters, that, if they had been set on the rack, they would never have confessed them, besides his poetical describing the circumstances of their meetings, as the well ordering of a banquet, the delicacy of a walk, with interlacing mere tales, as Gyges' ring and others, which who knoweth not to be flowers of poetry did never walk into Apollo's garden.

And even historiographers (although their lips sound of things done, and verity be written in their foreheads) have been glad to borrow both fashion and, perchance, weight of the poets. So 90 Herodotus entitled his History by the name of the nine Muses; and both he and all the rest that followed him either stale or usurped of poetry their passionate describing of passions, the many particularities of battles, which no man could affirm; or, if that be denied me, long orations put in the mouths of great kings and captains, which it is certain they never pronounced.

So that truly neither philosopher nor historiographer could at the first have entered into the gates of popular judgements, if they had not taken a great passport of poetry, which in all nations at this day where learning flourisheth not, is plain to be seen; in all which they 100 have some feeling of poetry.

In Turkey, besides their law-giving divines, they have no other writers but poets. In our neighbour country Ireland, where truly learning goeth very bare, yet are their poets held in a devout reverence. Even among the most barbarous and simple Indians where no writing is, yet have they their poets who make and sing songs, which they call *areytos*, both of their ancestors' deeds and praises of their gods: a sufficient probability that, if ever learning come among them, it must be by having their hard dull wits softened and sharpened with the sweet delights of poetry—for until 110 they find a pleasure in the exercises of the mind, great promises of much knowledge will little persuade them that know not the fruits of knowledge. In Wales, the true remnant of the ancient Britons, as there are good authorities to show the long time they had poets, which they called bards, so through all the conquests of Romans, Saxons, Danes and Normans, some of whom did seek to ruin all memory of learning from among them, yet do their poets even to this day last; so as it is not more notable in soon beginning than in long continuing.

But since the authors of most of our sciences were the Romans, and before them the Greeks, let us a little stand upon their 120 authorities, but even so far as to see what names they have given unto this now scorned skill.

Among the Romans a poet was called *vates*, which is as much as a diviner, foreseer, or prophet, as by his conjoined words *vaticinium* and *vaticinari* is manifest: so heavenly a title did that excellent people bestow upon this heart-ravishing knowledge. And so far were they carried into the admiration thereof, that they thought in the chanceable hitting upon any such verses great foretokens of their following fortunes were placed. Whereupon grew the word of *Sortes Virgilianae*, when by sudden opening Virgil's book they 130 lighted upon any verse of his making, whereof the histories of the emperors' lives are full: as of Albinus, the governor of our island, who in his childhood met with this verse

Arma amens capio nec sat rationis in armis

and in his age performed it. Which, although it were a very vain and godless superstition, as also it was to think spirits were commanded by such verses—whereupon this word charms, derived of *carmina*, cometh—so yet serveth it to show the great reverence those wits were held in; and altogether not without ground, since both the oracles of Delphos and Sibylla's prophecies were wholly 140 delivered in verses. For that same exquisite observing of number and measure in the words, and that high flying liberty of conceit proper to the poet, did seem to have some divine force in it.

And may not I presume a little further, to show the reasonableness of this word *vates*, and say that the holy David's Psalms are a divine poem? If I do, I shall not do it without the testimony of great learned men, both ancient and modern. But even the name of Psalms will speak for me, which being interpreted, is nothing but songs; then that it is fully written in metre, as all learned Hebricians agree, although the rules be not yet fully found; lastly and princi- 150 pally, his handling his prophecy, which is merely poetical: for what else is the awaking his musical instruments, the often and free changing of persons, his notable *prosopopoeias*, when he maketh you, as it were, see God coming in His majesty, his telling of the beasts' joyfulness and hills leaping, but a heavenly poesy, wherein almost he showeth himself a passionate lover of that unspeakable and everlasting beauty to be seen by the eyes of the mind, only cleared by faith? But truly now having named him, I fear me I seem to

profane that holy name, applying it to poetry, which is among us
160 thrown down to so ridiculous an estimation. But they that with quiet judgements will look a little deeper into it, shall find the end and working of it such as, being rightly applied, deserveth not to be scourged out of the Church of God.

But now let us see how the Greeks named it, and how they deemed of it. The Greeks called him a 'poet', which name hath, as the most excellent, gone through other languages. It cometh of this word *poiein*, which is, to make: wherein, I know not whether by luck or wisdom, we Englishmen have met with the Greeks in calling him a maker: which name, how high and incomparable a title it is, I had
170 rather were known by marking the scope of other sciences than by any partial allegation.

There is no art delivered to mankind that hath not the works of nature for his principal object, without which they could not consist, and on which they so depend, as they become actors and players, as it were, of what nature will have set forth. So doth the astronomer look upon the stars, and, by that he seeth, set down what order nature hath taken therein. So doth the geometrician and arithmetician in their diverse sorts of quantities. So doth the musician in times tell you which by nature agree, which not. The natural
180 philosopher thereon hath his name, and the moral philosopher standeth upon the natural virtues, vices, or passions of man; and follow nature (saith he) therein, and thou shalt not err. The lawyer saith what men have determined; the historian what men have done. The grammarian speaketh only of the rules of speech; and the rhetorician and logician, considering what in nature will soonest prove and persuade, thereon give artificial rules, which still are compassed within the circle of a question according to the proposed matter. The physician weigheth the nature of man's body, and the nature of things helpful or hurtful unto it. And the metaphysic,
190 though it be in the second and abstract notions, and therefore be counted supernatural, yet doth he indeed build upon the depth of nature. Only the poet, disdaining to be tied to any such subjection, lifted up with the vigour of his own invention, doth grow in effect another nature, in making things either better than nature bringeth forth, or, quite anew, forms such as never were in nature, as the Heroes, Demigods, Cyclops, Chimeras, Furies, and such like: so as he goeth hand in hand with nature, not enclosed within the narrow warrant of her gifts, but freely ranging only within the zodiac of his own wit. Nature never set forth the earth in so rich tapestry

as divers poets have done; neither with so pleasant rivers, fruitful trees, sweet-smelling flowers, nor whatsoever else may make the too much loved earth more lovely. Her world is brazen, the poets only deliver a golden. 200

But let those things alone, and go to man—for whom as the other things are, so it seemeth in him her uttermost cunning is employed —and know whether she have brought forth so true a lover as Theagenes, so constant a friend as Pylades, so valiant a man as Orlando, so right a prince as Xenophon's Cyrus, so excellent a man every way as Virgil's Aeneas. Neither let this be jestingly conceived, because the works of the one be essential, the other in imitation or fiction; for any understanding knoweth the skill of each artificer standeth in that *idea* or fore-conceit of the work, and not in the work itself. And that the poet hath that *idea* is manifest, by delivering them forth in such excellency as he had imagined them. Which delivering forth also is not wholly imaginative, as we are wont to say by them that build castles in the air; but so far substantially it worketh, not only to make a Cyrus, which had been but a particular excellency as nature might have done, but to bestow a Cyrus upon the world to make many Cyruses, if they will learn aright why and how that maker made him. 210 220

Neither let it be deemed too saucy a comparison to balance the highest point of man's wit with the efficacy of nature; but rather give right honour to the heavenly Maker of that maker, who having made man to His own likeness, set him beyond and over all the works of that second nature: which in nothing he showeth so much as in poetry, when with the force of a divine breath he bringeth things forth surpassing her doings—with no small arguments to the incredulous of that first accursed fall of Adam, since our erected wit maketh us know what perfection is, and yet our infected will keepeth us from reaching unto it. But these arguments will by few be understood, and by fewer granted. This much (I hope) will be given me, that the Greeks with some probability of reason gave him the name above all names of learning. 230

Now let us go to a more ordinary opening of him, that the truth may be the more palpable: and so I hope, though we get not so unmatched a praise as the etymology of his names will grant, yet his very description, which no man will deny, shall not justly be barred from a principal commendation.

Poesy therefore is an art of imitation, for so Aristotle termeth it in the word *mimesis*—that is to say, a representing, counterfeiting, 240 or figuring forth—to speak metaphorically, a speaking picture—with this end, to teach and delight.

Of this have been three general kinds. The chief, both in antiquity and excellency, were they that did imitate the unconceivable excellencies of God. Such were David in his Psalms; Solomon in his Song of Songs, in his Ecclesiastes, and Proverbs; Moses and Deborah in their Hymns; and the writer of Job: which, beside other, the learned Emanuel Tremellius and Franciscus Junius do entitle the poetical part of the Scripture. Against these none will speak that
250 hath the Holy Ghost in due holy reverence. (In this kind, though in a full wrong divinity, were Orpheus, Amphion, Homer in his Hymns, and many other, both Greeks and Romans.) And this poesy must be used by whosoever will follow St James's counsel in singing psalms when they are merry, and I know is used with the fruit of comfort by some, when, in sorrowful pangs of their death-bringing sins, they find the consolation of the never-leaving goodness.

The second kind is of them that deal with matters philosophical, either moral, as Tyrtaeus, Phocylides, Cato, or natural, as Lucretius and Virgil's *Georgics*; or astronomical, as Manilius and Pontanus; or
260 historical, as Lucan: which who mislike, the fault is in their judgement quite out of taste, and not in the sweet food of sweetly uttered knowledge.

But because this second sort is wrapped within the fold of the proposed subject, and takes not the course of his own invention, whether they properly be poets or no let grammarians dispute, and go to the third, indeed right poets, of whom chiefly this question ariseth: betwixt whom and these second is such a kind of difference as betwixt the meaner sort of painters, who counterfeit only such faces as are set before them, and the more excellent, who having
270 no law but wit, bestow that in colours upon you which is fittest for the eye to see: as the constant though lamenting look of Lucretia, when she punished in herself another's fault, wherein he painteth not Lucretia whom he never saw, but painteth the outward beauty of such a virtue. For these third be they which most properly do imitate to teach and delight, and to imitate borrow nothing of what is, hath been, or shall be; but range, only reined with learned discretion, into the divine consideration of what may be and should be. These be they that, as the first and most noble sort may justly be termed *vates*, so these are waited on in the excellentest languages
280 and best understandings with the fore-described name of poets. For these indeed do merely make to imitate, and imitate both to delight

and teach; and delight, to move men to take that goodness in hand, which without delight they would fly as from a stranger; and teach, to make them know that goodness whereunto they are moved—which being the noblest scope to which ever any learning was directed, yet want there not idle tongues to bark at them.

These be subdivided into sundry more special denominations. The most notable be the heroic, lyric, tragic, comic, satiric, iambic, elegiac, pastoral, and certain others, some of these being termed according to the matter they deal with, some by the sorts of verses 290 they liked best to write in; for indeed the greatest part of poets have apparelled their poetical inventions in that numbrous kind of writing which is called verse—indeed but apparelled, verse being but an ornament and no cause to poetry, since there have been many most excellent poets that never versified, and now swarm many versifiers that need never answer to the name of poets. For Xenophon, who did imitate so excellently as to give us *effigiem iusti imperii*, the portraiture of a just empire, under the name of Cyrus, (as Cicero saith of him) made therein an absolute heroical poem. So did Heliodorus in his sugared invention of that picture of love 300 in Theagenes and Chariclea; and yet both these wrote in prose: which I speak to show that it is not rhyming and versing that maketh a poet—no more than a long gown maketh an advocate, who though he pleaded in armour should be an advocate and no soldier. But it is that feigning notable images of virtues, vices, or what else, with that delightful teaching, which must be the right describing note to know a poet by; although indeed the senate of poets hath chosen verse as their fittest raiment, meaning, as in matter they passed all in all, so in manner to go beyond them: not speaking (table-talk fashion or like men in a dream) words as they chanceably fall from 310 the mouth, but peising each syllable of each word by just proportion according to the dignity of the subject.

Now therefore it shall not be amiss first to weigh this latter sort of poetry by his works, and then by his parts; and if in neither of these anatomies he be condemnable, I hope we shall obtain a more favourable sentence.

This purifying of wit—this enriching of memory, enabling of judgement, and enlarging of conceit—which commonly we call learning, under what name soever it come forth, or to what immediate end soever it be directed, the final end is to lead and 320 draw us to as high a perfection as our degenerate souls, made worse by their clayey lodgings, can be capable of.

This, according to the inclination of the man, bred many-formed impressions. For some that thought this felicity principally to be gotten by knowledge, and no knowledge to be so high or heavenly as acquaintance with the stars, gave themselves to astronomy; others, persuading themselves to be demigods if they knew the causes of things, became natural and supernatural philosophers; some an admirable delight drew to music; and some the certainty
330 of demonstration to the mathematics. But all, one and other, having this scope: to know, and by knowledge to lift up the mind from the dungeon of the body to the enjoying his own divine essence.

But when by the balance of experience it was found that the astronomer, looking to the stars, might fall in a ditch, that the inquiring philosopher might be blind in himself, and the mathematician might draw forth a straight line with a crooked heart, then lo, did proof, the overruler of opinions, make manifest that all these are but serving sciences, which, as they have each a private end in themselves, so yet are they all directed to the highest end of the
340 mistress-knowledge, by the Greeks called *architektonike*, which stands (as I think) in the knowledge of a man's self, in the ethic and politic consideration, with the end of well-doing and not of well-knowing only—even as the saddler's next end is to make a good saddle, but his further end to serve a nobler faculty, which is horsemanship, so the horseman's to soldiery, and the soldier not only to have the skill, but to perform the practice of a soldier. So that, the ending end of all earthly learning being virtuous action, those skills that most serve to bring forth that have a most just title to be princes over all the rest.

350 Wherein, if we can, show we the poet's nobleness, by setting him before his other competitors. Among whom as principal challengers step forth the moral philosophers, whom, me thinketh, I see coming towards me with a sullen gravity, as though they could not abide vice by daylight, rudely clothed for to witness outwardly their contempt of outward things, with books in their hands against glory, whereto they set their names, sophistically speaking against subtlety, and angry with any man in whom they see the foul fault of anger. These men casting largesse as they go, of definitions, divisions, and distinctions, with a scornful interrogative do soberly ask whether it
360 be possible to find any path so ready to lead a man to virtue as that which teacheth what virtue is; and teach it not only by delivering forth his very being, his causes and effects, but also by making known his enemy, vice, which must be destroyed, and his cumber-

some servant, passion, which must be mastered; by showing the generalities that containeth it, and the specialities that are derived from it; lastly, by plain setting down how it extendeth itself out of the limits of a man's own little world to the government of families and maintaining of public societies.

The historian scarcely giveth leisure to the moralist to say so much, but that he, laden with old mouse-eaten records, authorizing 370 himself (for the most part) upon other histories, whose greatest authorities are built upon the notable foundation of hearsay; having much ado to accord differing writers and to pick truth out of their partiality; better acquainted with a thousand years ago than with the present age, and yet better knowing how this world goeth than how his own wit runneth; curious for antiquities and inquisitive of novelties; a wonder to young folks and a tyrant in table talk, denieth, in a great chafe, that any man for teaching of virtue, and virtuous actions is comparable to him. 'I am *testis temporum, lux veritatis, vita memoriae, magistra vitae, nuntia vetustatis*. The philosopher', saith 380 he, 'teacheth a disputative virtue, but I do an active. His virtue is excellent in the dangerless Academy of Plato, but mine showeth forth her honourable face in the battles of Marathon, Pharsalia, Poitiers, and Agincourt. He teacheth virtue by certain abstract considerations, but I only bid you follow the footing of them that have gone before you. Old-aged experience goeth beyond the fine-witted philosopher, but I give the experience of many ages. Lastly, if he make the songbook, I put the learner's hand to the lute; and if he be the guide, I am the light.' Then would he allege you innumerable examples, confirming story by stories, how much the 390 wisest senators and princes have been directed by the credit of history, as Brutus, Alphonsus of Aragon, and who not, if need be? At length the long line of their disputation maketh a point in this, that the one giveth the precept, and the other the example.

Now whom shall we find (since the question standeth for the highest form in the school of learning) to be moderator? Truly, as me seemeth, the poet; and if not a moderator, even the man that ought to carry the title from them both, and much more from all other serving sciences. Therefore compare we the poet with the historian and with the moral philosopher; and if he go beyond them 400 both, no other human skill can match him. For as for the divine, with all reverence it is ever to be excepted, not only for having his scope as far beyond any of these as eternity exceedeth a moment, but even for passing each of these in themselves. And for the

lawyer, though *Ius* be the daughter of Justice, and justice the chief of virtues, yet because he seeketh to make men good rather *formidine poenae* than *virtutis amore*; or, to say righter, doth not endeavour to make men good, but that their evil hurt not others; having no care, so he be a good citizen, how bad a man he be:
410 therefore as our wickedness maketh him necessary, and necessity maketh him honourable, so is he not in the deepest truth to stand in rank with these who all endeavour to take naughtiness away and plant goodness even in the secretest cabinet of our souls. And these four are all that any way deal in that consideration of men's manners, which being the supreme knowledge, they that best breed it deserve the best commendation.

The philosopher, therefore, and the historian are they which would win the goal, the one by precept, the other by example. But both, not having both, do both halt. For the philosopher, setting
420 down with thorny arguments the bare rule, is so hard of utterance and so misty to be conceived, that one that hath no other guide but him shall wade in him till he be old before he shall find sufficient cause to be honest. For his knowledge standeth so upon the abstract and general, that happy is that man who may understand him, and more happy that can apply what he doth understand. On the other side, the historian, wanting the precept, is so tied, not to what should be but to what is, to the particular truth of things and not to the general reason of things, that his example draweth no necessary consequence, and therefore a less fruitful doctrine.

430 Now doth the peerless poet perform both: for whatsoever the philosopher saith should be done, he giveth a perfect picture of it in someone by whom he presupposeth it was done, so as he coupleth the general notion with the particular example. A perfect picture I say, for he yieldeth to the powers of the mind an image of that whereof the philosopher bestoweth but a wordish description, which doth neither strike, pierce, nor possess the sight of the soul so much as that other doth. For as in outward things, to a man that had never seen an elephant or a rhinoceros, who should tell him most exquisitely all their shapes, colour, bigness, and particular
440 marks, or of a gorgeous palace the architecture, with declaring the full beauties, might well make the hearer able to repeat, as it were by rote, all he had heard, yet should never satisfy his inward conceit with being witness to itself of a true lively knowledge; but the same man, as soon as he might see those beasts well painted, or that house well in model, should straightways grow, without need of any

description, to a judicial comprehending of them: so no doubt the philosopher with his learned definitions—be it of virtue, vices, matters of public policy or private government—replenisheth the memory with many infallible grounds of wisdom, which, notwithstanding, lie dark before the imaginative and judging power, if they 450 be not illuminated or figured forth by the speaking picture of poesy.

Tully taketh much pains, and many times not without poetical helps, to make us know the force love of our country hath in us. Let us but hear old Anchises speaking in the midst of Troy's flames, or see Ulysses in the fulness of all Calypso's delights bewail his absence from barren and beggarly Ithaca. Anger, the Stoics said, was a short madness: let but Sophocles bring you Ajax on a stage, killing or whipping sheep and oxen, thinking them the army of Greeks, with their chieftains Agamemnon and Menelaus, and tell me if you have not a more familiar insight into anger than finding 460 in the schoolmen his *genus* and difference. See whether wisdom and temperance in Ulysses and Diomedes, valour in Achilles, friendship in Nisus and Euryalus, even to an ignorant man carry not an apparent shining; and, contrarily, the remorse of conscience in Oedipus, the soon repenting pride in Agamemnon, the self-devouring cruelty in his father Atreus, the violence of ambition in the two Theban brothers, the sour-sweetness of revenge in Medea; and, to fall lower, the Terentian Gnatho and our Chaucer's Pandar so expressed that we now use their names to signify their trades: and finally, all virtues, vices, and passions so in their own natural 470 seats laid to the view, that we seem not to hear of them, but clearly to see through them.

But even in the most excellent determination of goodness, what philosopher's counsel can so readily direct a prince, as the feigned Cyrus in Xenophon; or a virtuous man in all fortunes, as Aeneas in Virgil; or a whole commonwealth, as the way of Sir Thomas More's *Utopia*? I say the way, because where Sir Thomas More erred, it was the fault of the man and not of the poet, for that way of patterning a commonwealth was most absolute, though he perchance hath not so absolutely performed it. For the question is, 480 whether the feigned image of poetry or the regular instruction of philosophy hath the more force in teaching: wherein if the philosophers have more rightly showed themselves philosophers than the poets have attained to the high top of their profession, as in truth

Mediocribus esse poetis,
Non dii, non homines, non concessere columnae;

it is, I say again, not the fault of the art, but that by few men that art can be accomplished.

Certainly, even our Saviour Christ could as well have given the 490 moral commonplaces of uncharitableness and humbleness as the divine narration of Dives and Lazarus; or of disobedience and mercy, as that heavenly discourse of the lost child and the gracious father; but that His through-searching wisdom knew the estate of Dives burning in hell, and of Lazarus in Abraham's bosom, would more constantly (as it were) inhabit both the memory and judgement. Truly, for myself, meseems I see before mine eyes the lost child's disdainful prodigality, turned to envy a swine's dinner: which by the learned divines are thought not historical acts, but instructing parables.

500 For conclusion, I say the philosopher teacheth, but he teacheth obscurely, so as the learned only can understand him, that is to say, he teacheth them that are already taught; but the poet is the food for the tenderest stomachs, the poet is indeed the right popular philosopher, whereof Aesop's tales give good proof: whose pretty allegories, stealing under the formal tales of beasts, make many, more beastly than beasts, begin to hear the sound of virtue from these dumb speakers.

But now may it be alleged that if this imagining of matters be so fit for the imagination, then must the historian needs surpass, who 510 bringeth you images of true matters, such as indeed were done, and not such as fantastically or falsely may be suggested to have been done. Truly, Aristotle himself, in his discourse of poesy, plainly determineth this question, saying that poetry is *philosophoteron* and *spoudaioteron*, that is to say, it is more philosophical and more studiously serious than history. His reason is, because poesy dealeth with *katholou*, that is to say, with the universal consideration, and the history with *kathekaston*, the particular: now, saith he, the universal weighs what is fit to be said or done, either in likelihood or necessity (which the poesy considereth in his imposed names), 520 and the particular only marks whether Alcibiades did, or suffered, this or that. Thus far Aristotle: which reason of his (as all his) is most full of reason. For indeed, if the question were whether it were better to have a particular act truly or falsely set down, there is no doubt which is to be chosen, no more than whether you had rather have Vespasian's picture right as he was, or, at the painter's pleasure, nothing resembling. But if the question be for your own use and learning, whether it be better to have it set down as it should be, or as it was, then certainly is more doctrinable the feigned Cyrus in Xenophon than the true Cyrus in Justin, and the feigned Aeneas in Virgil than the right Aeneas in Dares Phrygius: 530 as to a lady that desired to fashion her countenance to the best grace, a painter should more benefit her to portrait a most sweet face, writing Canidia upon it, than to paint Canidia as she was, who, Horace sweareth, was full ill-favoured.

If the poet do his part aright, he will show you in Tantalus, Atreus, and such like, nothing that is not to be shunned; in Cyrus, Aeneas, Ulysses, each thing to be followed; where the historian, bound to tell things as things were, cannot be liberal (without he will be poetical) of a perfect pattern, but, as in Alexander or Scipio himself, show doings, some to be liked, some to be misliked. And 540 then how will you discern what to follow but by your own discretion, which you had without reading Quintus Curtius? And whereas a man may say, though in universal consideration of doctrine the poet prevaileth, yet that the history, in his saying such a thing was done, doth warrant a man more in that he shall follow—the answer is manifest: that, if he stand upon that was (as if he should argue, because it rained yesterday, therefore it should rain today), then indeed hath it some advantage to a gross conceit; but if he know an example only informs a conjectured likelihood, and so go by reason, the poet doth so far exceed him as he is to frame his 550 example to that which is most reasonable (be it in warlike, politic, or private matters), where the historian in his bare 'was' hath many times that which we call fortune to overrule the best wisdom. Many times he must tell events whereof he can yield no cause; or, if he do, it must be poetically.

For that a feigned example hath as much force to teach as a true example (for as for to move, it is clear, since the feigned may be tuned to the highest key of passion), let us take one example wherein an historian and a poet did concur. Herodotus and Justin do both testify that Zopyrus, King Darius' faithful servant, seeing 560 his master long resisted by the rebellious Babylonians, feigned himself in extreme disgrace of his king: for verifying of which, he caused his own nose and ears to be cut off, and so flying to the Babylonians, was received, and for his known valour so sure credited, that he did find means to deliver them over to Darius. Much like matter doth Livy record of Tarquinius and his son. Xenophon excellently feigneth such another stratagem performed by Abradatas in Cyrus' behalf. Now would I fain know, if occasion

be presented unto you to serve your prince by such an honest
570 dissimulation, why you do not as well learn it of Xenophon's fiction
as of the other's verity; and truly so much the better, as you shall
save your nose by the bargain: for Abradatas did not counterfeit so
far. So then the best of the historian is subject to the poet; for
whatsoever action, or faction, whatsoever counsel, policy, or war
stratagem the historian is bound to recite, that may the poet (if he
list) with his imitation make his own, beautifying it both for further
teaching, and more delighting, as it please him: having all, from
Dante's heaven to his hell, under the authority of his pen. Which
if I be asked what poets have done so, as I might well name some,
580 so yet say I, and say again, I speak of the art, and not of the artificer.

Now, to that which commonly is attributed to the praise of
history, in respect of the notable learning is got by marking the
success, as though therein a man should see virtue exalted and vice
punished—truly that commendation is particular to poetry, and far
off from history. For indeed poetry ever sets virtue so out in her
best colours, making Fortune her well-waiting handmaid, that one
must needs be enamoured of her. Well may you see Ulysses in a
storm, and in other hard plights; but they are but exercises of
patience and magnanimity, to make them shine the more in the
590 near-following prosperity. And of the contrary part, if evil men
come to the stage, they ever go out (as the tragedy writer answered
to one that misliked the show of such persons) so manacled as they
little animate folks to follow them. But the history, being captived
to the truth of a foolish world, is many times a terror from well-
doing, and an encouragement to unbridled wickedness. For see we
not valiant Miltiades rot in his fetters? The just Phocion and the
accomplished Socrates put to death like traitors? The cruel Severus
live prosperously? The excellent Severus miserably murdered?
Sulla and Marius dying in their beds? Pompey and Cicero slain
600 then when they would have thought exile a happiness? See we not
virtuous Cato driven to kill himself, and rebel Caesar so advanced
that his name yet, after 1600 years, lasteth in the highest honour?
And mark but even Caesar's own words of the aforenamed Sulla
(who in that only did honestly, to put down his dishonest tyranny),
literas nescivit, as if want of learning caused him to do well. He
meant it not by poetry, which, not content with earthly plagues,
deviseth new punishments in hell for tyrants, nor yet by philosophy,
which teacheth *occidendos esse*; but no doubt by skill in history,
for that indeed can afford you Cypselus, Periander, Phalaris,

Dionysius, and I know not how many more of the same kennel, that speed well enough in their abominable injustice of usurpation. 610

I conclude, therefore, that he excelleth history, not only in furnishing the mind with knowledge, but in setting it forward to that which deserveth to be called and accounted good: which setting forward, and moving to well-doing, indeed setteth the laurel crown upon the poets as victorious, not only of the historian, but over the philosopher, howsoever in teaching it may be questionable.

For suppose it be granted (that which I suppose with great reason may be denied) that the philosopher, in respect of his methodical proceeding, doth teach more perfectly than the poet, yet do I think 620 that no man is so much *philophilosophos* as to compare the philosopher in moving with the poet. And that moving is of a higher degree than teaching, it may by this appear, that it is well nigh both the cause and effect of teaching. For who will be taught, if he be not moved with desire to be taught? And what so much good doth that teaching bring forth (I speak still of moral doctrine) as that it moveth one to do that which it doth teach? For, as Aristotle saith, it is not *gnosis* but *praxis* must be the fruit. And how *praxis* can be, without being moved to practise, it is no hard matter to consider.

The philosopher showeth you the way, he informeth you of the 630 particularities, as well of the tediousness of the way, as of the pleasant lodging you shall have when your journey is ended, as of the many by-turnings that may divert you from your way. But this is to no man but to him that will read him, and read him with attentive studious painfulness; which constant desire whosoever hath in him, hath already passed half the hardness of the way, and therefore is beholding to the philosopher but for the other half. Nay truly, learned men have learnedly thought that where once reason hath so much overmastered passion as that the mind hath a free desire to do well, the inward light each mind hath in itself is as 640 good as a philosopher's book; since in nature we know it is well to do well, and what is well, and what is evil, although not in the words of art which philosophers bestow upon us; for out of natural conceit the philosophers drew it. But to be moved to do that which we know, or to be moved with desire to know: *hoc opus, hic labor est.*

Now therein of all sciences (I speak still of human, and according to the human conceit) is our poet the monarch. For he doth not only show the way, but giveth so sweet a prospect into the way, as will entice any man to enter into it. Nay, he doth, as if your journey should lie through a fair vineyard, at the first give you a cluster of 650

grapes, that full of that taste, you may long to pass further. He beginneth not with obscure definitions, which must blur the margin with interpretations, and load the memory with doubtfulness; but he cometh to you with words set in delightful proportion, either accompanied with, or prepared for, the well enchanting skill of music; and with a tale forsooth he cometh unto you, with a tale which holdeth children from play, and old men from the chimney corner. And, pretending no more, doth intend the winning of the mind from wickedness to virtue—even as the child is often brought
660 to take most wholesome things by hiding them in such other as have a pleasant taste, which, if one should begin to tell them the nature of aloes or rhubarbum they should receive, would sooner take their physic at their ears than at their mouth. So is it in men (most of which are childish in the best things, till they be cradled in their graves): glad will they be to hear the tales of Hercules, Achilles, Cyrus, Aeneas; and, hearing them, must needs hear the right description of wisdom, valour, and justice; which, if they had been barely, that is to say philosophically, set out, they would swear they be brought to school again.

670 That imitation whereof poetry is, hath the most conveniency to nature of all other, insomuch that, as Aristotle saith, those things which in themselves are horrible, as cruel battles, unnatural monsters, are made in poetical imitation delightful. Truly, I have known men that even with reading *Amadis de Gaule* (which God knoweth wanteth much of a perfect poesy) have found their hearts moved to the exercise of courtesy, liberality, and especially courage. Who readeth Aeneas carrying old Anchises on his back, that wisheth not it were his fortune to perform so excellent an act? Whom doth not these words of Turnus move, the tale of Turnus
680 having planted his image in the imagination,

Fugientem haec terra videbit?
Usque adeone mori miserum est?

Where the philosophers, as they scorn to delight, so much they be content little to move—saving wrangling whether *virtus* be the chief or the only good, whether the contemplative or the active life do excel—which Plato and Boethius well knew, and therefore made mistress Philosophy very often borrow the masking raiment of poesy. For even those hard-hearted evil men who think virtue a school name, and know no other good but *indulgere genio*, and
690 therefore despise the austere admonitions of the philosopher, and

feel not the inward reason they stand upon, yet will be content to be delighted—which is all the good-fellow poet seemeth to promise—and so steal to see the form of goodness (which seen they cannot but love) ere themselves be aware, as if they took a medicine of cherries.

Infinite proofs of the strange effects of this poetical invention might be alleged; only two shall serve, which are so often remembered as I think all men know them. The one of Menenius Agrippa, who, when the whole people of Rome had resolutely divided themselves from the senate, with apparent show of utter ruin, 700 though he were (for that time) an excellent orator, came not among them upon trust of figurative speeches or cunning insinuations, and much less with far-fet maxims of philosophy, which (especially if they were Platonic) they must have learned geometry before they could well have conceived; but forsooth he behaves himself like a homely and familiar poet. He telleth them a tale, that there was a time when all the parts of the body made a mutinous conspiracy against the belly, which they thought devoured the fruits of each other's labour; they concluded they would let so unprofitable a spender starve. In the end, to be short (for the tale is notorious, 710 and as notorious that it was a tale), with punishing the belly they plagued themselves. This applied by him wrought such effect in the people, as I never read that only words brought forth but then so sudden and so good an alteration; for upon reasonable conditions a perfect reconcilement ensued. The other is of Nathan the prophet, who, when the holy David had so far forsaken God as to confirm adultery with murder, when he was to do the tenderest office of a friend in laying his own shame before his eyes, sent by God to call again so chosen a servant, how doth he it but by telling of a man whose beloved lamb was ungratefully taken from his 720 bosom? The application most divinely true, but the discourse itself feigned; which made David (I speak of the second and instrumental cause) as in a glass see his own filthiness, as that heavenly psalm of mercy well testifieth.

By these, therefore, examples and reasons, I think it may be manifest that the poet, with that same hand of delight, doth draw the mind more effectually than any other art doth. And so a conclusion not unfitly ensueth: that, as virtue is the most excellent resting place for all worldly learning to make his end of, so poetry, being the most familiar to teach it, and most princely to move towards it, 730 in the most excellent work is the most excellent workman.

But I am content not only to decipher him by his works (although works, in commendation or dispraise, must ever hold a high authority), but more narrowly will examine his parts; so that (as in a man) though all together may carry a presence full of majesty and beauty, perchance in some one defectuous piece we may find blemish.

Now in his parts, kinds, or species (as you list to term them), it is to be noted that some poesies have coupled together two or three 740 kinds, as the tragical and comical, whereupon is risen the tragi-comical. Some, in the manner, have mingled prose and verse, as Sannazaro and Boethius. Some have mingled matters heroical and pastoral. But that cometh all to one in this question, for, if severed they be good, the conjunction cannot be hurtful. Therefore, perchance forgetting some and leaving some as needless to be remembered, it shall not be amiss in a word to cite the special kinds, to see what faults may be found in the right use of them.

Is it then the Pastoral poem which is misliked? (For perchance where the hedge is lowest they will soonest leap over.) Is the poor 750 pipe disdained, which sometime out of Meliboeus' mouth can show the misery of people under hard lords or ravening soldiers? And again, by Tityrus, what blessedness is derived to them that lie lowest from the goodness of them that sit highest; sometimes, under the pretty tales of wolves and sheep, can include the whole considerations of wrong-doing and patience; sometimes show that contentions for trifles can get but a trifling victory: where perchance a man may see that even Alexander and Darius, when they strave who should be cock of this world's dunghill, the benefit they got was that the after-livers may say

760 Haec memini et victum frustra contendere Thyrsin:
Ex illo Corydon, Corydon est tempore nobis.

Or is it the lamenting Elegiac; which in a kind heart would move rather pity than blame; who bewails with the great philosopher Heraclitus, the weakness of mankind and the wretchedness of the world; who surely is to be praised, either for compassionate accompanying just causes of lamentations, or for rightly painting out how weak be the passions of woefulness? Is it the bitter but wholesome Iambic, who rubs the galled mind, in making shame the trumpet of villainy, with bold and open crying out against 770 naughtiness? Or the Satiric, who

Omne vafer vitium ridenti tangit amico;

who sportingly never leaveth till he make a man laugh at folly, and at length ashamed, to laugh at himself, which he cannot avoid without avoiding the folly; who, while

circum praecordia ludit,

giveth us to feel how many headaches a passionate life bringeth us to; how, when all is done,

Est Ulubris, animus si nos non deficit aequus?

No, perchance it is the Comic, whom naughty play-makers and stage-keepers have justly made odious. To the arguments of abuse I will answer after. Only this much now is to be said, that the comedy is an imitation of the common errors of our life, which he representeth in the most ridiculous and scornful sort that may be, so as it is impossible that any beholder can be content to be such a one. Now, as in geometry the oblique must be known as well as the right, and in arithmetic the odd as well as the even, so in the actions of our life who seeth not the filthiness of evil wanteth a great foil to perceive the beauty of virtue. This doth the comedy handle so in our private and domestical matters as with hearing it we get as it were an experience what is to be looked for of a niggardly Demea, of a crafty Davus, of a flattering Gnatho, of a vainglorious Thraso; and not only to know what effects are to be expected, but to know who be such, by the signifying badge given them by the comedian. And little reason hath any man to say that men learn the evil by seeing it so set out, since, as I said before, there is no man living but, by the force truth hath in nature, no sooner seeth these men play their parts, but wisheth them *in pistrinum*; although perchance the sack of his own faults lie so hidden behind his back that he seeth not himself dance the same measure; whereto yet nothing can more open his eyes than to find his own actions contemptibly set forth.

So that the right use of comedy will (I think) by nobody be blamed; and much less of the high and excellent Tragedy, that openeth the greatest wounds, and showeth forth the ulcers that are covered with tissue; that maketh kings fear to be tyrants, and tyrants manifest their tyrannical humours; that, with stirring the affects of admiration and commiseration, teacheth the uncertainty of this world, and upon how weak foundations gilden roofs are builded; that maketh us know

810 Qui sceptra saevus duro imperio regit
Timet timentes; metus in auctorem redit.

But how much it can move, Plutarch yieldeth a notable testimony of the abominable tyrant Alexander Pheraeus, from whose eyes a tragedy, well made and represented, drew abundance of tears, who without all pity had murdered infinite numbers, and some of his own blood: so as he, that was not ashamed to make matters for tragedies, yet could not resist the sweet violence of a tragedy. And if it wrought no further good in him, it was that he, in despite of himself, withdrew himself from hearkening to that which might 820 mollify his hardened heart. But it is not the tragedy they do mislike; for it were too absurd to cast out so excellent a representation of whatsoever is most worthy to be learned.

Is it the Lyric that most displeaseth, who with his tuned lyre and well-accorded voice, giveth praise, the reward of virtue, to virtuous acts; who gives moral precepts, and natural problems; who sometimes raiseth up his voice to the height of the heavens, in singing the lauds of the immortal God? Certainly, I must confess my own barbarousness, I never heard the old song of Percy and Douglas that I found not my heart moved more than with a trumpet; and 830 yet is it sung but by some blind crowder, with no rougher voice than rude style; which, being so evil apparelled in the dust and cobwebs of that uncivil age, what would it work trimmed in the gorgeous eloquence of Pindar? In Hungary I have seen it the manner at all feasts, and other such meetings, to have songs of their ancestors' valour, which that right soldierlike nation think one of the chiefest kindlers of brave courage. The incomparable Lacedemonians did not only carry that kind of music ever with them to the field, but even at home, as such songs were made, so were they all content to be singers of them—when the lusty men were 840 to tell what they did, the old men what they had done, and the young what they would do. And where a man may say that Pindar many times praiseth highly victories of small moment, matters rather of sport than virtue; as it may be answered, it was the fault of the poet, and not of the poetry, so indeed the chief fault was in the time and custom of the Greeks, who set those toys at so high a price that Philip of Macedon reckoned a horserace won at Olympus among his three fearful felicities. But as the unimitable

Pindar often did, so is that kind most capable and most fit to awake the thoughts from the sleep of idleness to embrace honourable enterprises. 850

There rests the Heroical—whose very name (I think) should daunt all backbiters: for by what conceit can a tongue be directed to speak evil of that which draweth with him no less champions than Achilles, Cyrus, Aeneas, Turnus, Tydeus, and Rinaldo?—who doth not only teach and move to a truth, but teacheth and moveth to the most high and excellent truth; who maketh magnanimity and justice shine through all misty fearfulness and foggy desires; who, if the saying of Plato and Tully be true, that who could see virtue would be wonderfully ravished with the love of her beauty—this man sets her out to make her more lovely in her holiday apparel, to the eye 860 of any that will deign not to disdain until they understand. But if anything be already said in the defence of sweet poetry, all concurreth to the maintaining the heroical, which is not only a kind, but the best and most accomplished kind of poetry. For as the image of each action stirreth and instructeth the mind, so the lofty image of such worthies most inflameth the mind with desire to be worthy, and informs with counsel how to be worthy. Only let Aeneas be worn in the tablet of your memory, how he governeth himself in the ruin of his country; in the preserving his old father, and carrying away his religious ceremonies; in obeying God's 870 commandment to leave Dido, though not only all passionate kindness, but even the human consideration of virtuous gratefulness, would have craved other of him; how in storms, how in sports, how in war, how in peace, how a fugitive, how victorious, how besieged, how besieging, how to strangers, how to allies, how to enemies, how to his own; lastly, how in his inward self, and how in his outward government—and I think, in a mind not prejudiced with a prejudicating humour, he will be found in excellency fruitful, yea, even as Horace saith,

melius Chrysippo et Crantore. 880

But truly I imagine it falleth out with these poet-whippers, as with some good women, who often are sick, but in faith they cannot tell where; so the name of poetry is odious to them, but neither his cause nor effects, neither the sum that contains him, nor the particularities descending from him, give any fast handle to their carping dispraise.

Since then poetry is of all human learning the most ancient and of most fatherly antiquity, as from whence other learnings have taken their beginnings; since it is so universal that no learned nation 890 doth despise it, nor barbarous nation is without it; since both Roman and Greek gave such divine names unto it, the one of prophesying, the other of making, and that indeed that name of making is fit for him, considering that where all other arts retain themselves within their subject, and receive, as it were, their being from it, the poet only bringeth his own stuff, and doth not learn a conceit out of a matter, but maketh matter for a conceit; since neither his description nor end containing any evil, the thing described cannot be evil; since his effects be so good as to teach goodness and to delight the learners; since therein (namely in moral 900 doctrine, the chief of all knowledges) he doth not only far pass the historian, but, for instructing, is well nigh comparable to the philosopher, for moving leaves him behind him; since the Holy Scripture (wherein there is no uncleanness) hath whole parts in it poetical, and that even our Saviour Christ vouchsafed to use the flowers of it; since all his kinds are not only in their united forms but in their severed dissections fully commendable; I think (and think I think rightly) the laurel crown appointed for triumphant captains doth worthily (of all other learnings) honour the poet's triumph.

910 But because we have ears as well as tongues, and that the lightest reasons that may be will seem to weigh greatly, if nothing be put in the counterbalance, let us hear, and, as well as we can, ponder what objections be made against this art, which may be worthy either of yielding or answering.

First, truly I note not only in these *misomousoi*, poet-haters, but in all that kind of people who seek a praise by dispraising others, that they do prodigally spend a great many wandering words in quips and scoffs, carping and taunting at each thing which, by stirring the spleen, may stay the brain from a through-beholding 920 the worthiness of the subject. Those kind of objections, as they are full of a very idle easiness, since there is nothing of so sacred a majesty but that an itching tongue may rub itself upon it, so deserve they no other answer, but, instead of laughing at the jest, to laugh at the jester. We know a playing wit can praise the discretion of an ass, the comfortableness of being in debt, and the jolly commodities of being sick of the plague. So of the contrary side, if we will turn Ovid's verse

Ut lateat virtus proximitate mali,

that good lie hid in nearness of the evil, Agrippa will be as merry in showing the vanity of science as Erasmus was in the commending of folly. Neither shall any man or matter escape some touch of these smiling railers. But for Erasmus and Agrippa, they had another foundation than the superficial part would promise. Marry, these other pleasant faultfinders, who will correct the verb before they understand the noun, and confute others' knowledge before they confirm their own—I would have them only remember that scoffing cometh not of wisdom. So as the best title in true English they get with their merriments is to be called good fools; for so have our grave forefathers ever termed that humorous kind of jesters. 930

But that which giveth greatest scope to their scorning humour is rhyming and versing. It is already said (and, as I think, truly said), it is not rhyming and versing that maketh poesy. One may be a poet without versing, and a versifier without poetry. But yet, presuppose it were inseparable (as indeed it seemeth Scaliger judgeth), truly it were an inseparable commendation. For if *oratio* next to *ratio*, speech next to reason, be the greatest gift bestowed upon mortality, that cannot be praiseless which doth most polish that blessing of speech; which considers each word, not only (as a man may say) by his most forcible quality, but by his best measured quantity, carrying even in themselves a harmony—without, perchance, number, measure, order, proportion be in our time grown odious. But lay aside the just praise it hath, by being the only fit speech for music (music, I say, the most divine striker of the senses), thus much is undoubtedly true, that if reading be foolish without remembering, memory being the only treasure of knowledge, those words which are fittest for memory are likewise most convenient for knowledge. Now, that verse far exceedeth prose in the knitting up of memory, the reason is manifest: the words (besides their delight, which hath a great affinity to memory) being so set as one cannot be lost but the whole work fails; which accusing itself, calleth the remembrance back to itself, and so most strongly confirmeth it. Besides, one word so, as it were, begetting another, as, be it in rhyme or measured verse, by the former a man shall have a near guess to the follower. Lastly, even they that have taught the art of memory have showed nothing so apt for it as a certain room divided into many places well and thoroughly known. Now, that hath the verse in effect perfectly, every word having his natural seat, which seat must needs make the 940 950 960

word remembered. But what needeth more in a thing so known to all men? Who is it that ever was a scholar that doth not carry away
970 some verses of Virgil, Horace, or Cato, which in his youth he learned, and even to his old age serve him for hourly lessons? But the fitness it hath for memory is notably proved by all delivery of arts: wherein for the most part, from grammar to logic, mathematics, physic, and the rest, the rules chiefly necessary to be borne away are compiled in verses. So that, verse being in itself sweet and orderly, and being best for memory, the only handle of knowledge, it must be in jest that any man can speak against it.

Now then go we to the most important imputations laid to the poor poets. For aught I can yet learn, they are these. First, that
980 there being many other more fruitful knowledges, a man might better spend his time in them than in this. Secondly, that it is the mother of lies. Thirdly, that it is the nurse of abuse, infecting us with many pestilent desires; with a siren's sweetness drawing the mind to the serpent's tail of sinful fancies (and herein, especially, comedies give the largest field to ear, as Chaucer saith); how, both in other nations and in ours, before poets did soften us, we were full of courage, given to martial exercises, the pillars of manlike liberty, and not lulled asleep in shady idleness with poets' pastimes. And lastly, and chiefly, they cry out with open mouth as if they
990 had overshot Robin Hood, that Plato banished them out of his commonwealth. Truly, this is much, if there be much truth in it.

First, to the first. That a man might better spend his time, is a reason indeed; but it doth (as they say) but *petere principium*. For if it be as I affirm, that no learning is so good as that which teacheth and moveth to virtue; and none can both teach and move thereto so much as poetry: then is the conclusion manifest that ink and paper cannot be to a more profitable purpose employed. And certainly, though a man should grant their first assumption, it should follow (methinks) very unwillingly, that good is not good,
1000 because better is better. But I still and utterly deny that there is sprong out of earth a more fruitful knowledge.

To the second, therefore, that they should be the principal liars, I will answer paradoxically, but truly, I think truly, that of all writers under the sun the poet is the least liar, and, though he would, as a poet can scarcely be a liar. The astronomer, with his cousin the geometrician, can hardly escape, when they take upon them to measure the height of the stars. How often, think you, do the physicians lie, when they aver things good for sicknesses, which afterwards send Charon a great number of souls drowned in a potion before they come to his ferry? And no less of the rest, which take upon them to affirm. Now, for the poet, he nothing affirms, and therefore never lieth. For, as I take it, to lie is to affirm that to be true which is false. So as the other artists, and especially the historian, affirming many things, can, in the cloudy knowledge of mankind, hardly escape from many lies. But the poet (as I said before) never affirmeth. The poet never maketh any circles about your imagination, to conjure you to believe for true what he writes. He citeth not authorities of other histories, but even for his entry calleth the sweet Muses to inspire into him a good invention; in truth, not labouring to tell you what is or is not, but what should or should not be. And therefore, though he recount things not true, yet because he telleth them not for true, he lieth not—without we will say that Nathan lied in his speech before-alleged to David; which as a wicked man durst scarce say, so think I none so simple would say Aesop lied in the tales of his beasts; for who thinks that Aesop wrote it for actually true were well worthy to have his name chronicled among the beasts he writeth of. What child is there, that, coming to a play, and seeing *Thebes* written in great letters upon an old door, doth believe that it is Thebes? If then a man can arrive to that child's age to know that the poets' persons and doings are but pictures what should be, and not stories what have been, they will never give the lie to things not affirmatively but allegorically and figuratively written. And therefore, as in history, looking for truth, they may go away full fraught with falsehood, so in poesy, looking but for fiction, they shall use the narration but as an imaginative ground-plot of a profitable invention. But hereto is replied, that the poets give names to men they write of, which argueth a conceit of an actual truth, and so, not being true, proves a falsehood. And doth the lawyer lie then, when under the names of *John-a-stiles* and *John-a-nokes* he puts his case? But that is easily answered. Their naming of men is but to make their picture the more lively, and not to build any history: painting men, they cannot leave men nameless. We see we cannot play at chess but that we must give names to our chessmen; and yet, methinks, he were a very partial champion of truth that would say we lied for giving a piece of wood the reverend title of a bishop. The poet nameth Cyrus or Aeneas no other way than to show what men of their fames, fortunes, and estates should do.

Their third is, how much it abuseth men's wit, training it to wanton sinfulness and lustful love: for indeed that is the principal, if not only, abuse I can hear alleged. They say, the comedies rather teach than reprehend amorous conceits. They say the lyric is larded with passionate sonnets; the elegiac weeps the want of his mistress; and that even to the heroical, Cupid hath ambitiously climbed. Alas, Love, I would thou couldst as well defend thyself as thou canst offend others. I would those on whom thou dost attend could either put thee away, or yield good reason why they keep thee. But grant love of beauty to be a beastly fault (although it be very hard, since only man, and no beast, hath that gift to discern beauty); grant that lovely name of Love to deserve all hateful reproaches (although even some of my masters the philosophers spent a good deal of their lamp-oil in setting forth the excellency of it); grant, I say, whatsoever they will have granted, that not only love, but lust, but vanity, but (if they list) scurrility, possesseth many leaves of the poets' books; yet think I, when this is granted, they will find their sentence may with good manners put the last words foremost, and not say that poetry abuseth man's wit, but that man's wit abuseth poetry.

For I will not deny but that man's wit may make poesy, which should be *cikastiké* (which some learned have defined: figuring forth good things), to be *phantastiké* (which doth, contrariwise, infect the fancy with unworthy objects), as the painter, that should give to the eye either some excellent perspective, or some fine picture, fit for building or fortification, or containing in it some notable example (as Abraham sacrificing his son Isaac, Judith killing Holofernes, David fighting with Goliath), may leave those, and please an ill-pleased eye with wanton shows of better hidden matters. But what, shall the abuse of a thing make the right use odious? Nay truly, though I yield that poesy may not only be abused, but that being abused, by the reason of his sweet charming force, it can do more hurt than any other army of words: yet shall it be so far from concluding that the abuse should give reproach to the abused, that, contrariwise, it is a good reason that whatsoever, being abused, doth most harm, being rightly used (and upon the right use each thing conceiveth his title), doth most good. Do we not see the skill of physic, the best rampire to our often-assaulted bodies, being abused, teach poison, the most violent destroyer? Doth not knowledge of law, whose end is to even and right all things, being abused, grow the crooked fosterer of horrible injuries? Doth not (to go to the highest) God's word abused breed heresy, and His name abused become blasphemy? Truly, a needle cannot do much hurt, and as truly (with leave of ladies be it spoken) it cannot do much good: with a sword thou may'st kill thy father, and with a sword thou may'st defend thy prince and country. So that, as in their calling poets fathers of lies they said nothing, so in this their argument of abuse they prove the commendation.

They allege herewith, that before poets began to be in price our nation had set their hearts' delight upon action, and not imagination: rather doing things worthy to be written, than writing things fit to be done. What that before-time was, I think scarcely Sphinx can tell, since no memory is so ancient that hath not the precedent of poetry. And certain it is that, in our plainest homeliness, yet never was the Albion nation without poetry. Marry, this argument, though it be levelled against poetry, yet is it indeed a chainshot against all learning, or bookishness as they commonly term it. Of such mind were certain Goths, of whom it is written that, having in the spoil of a famous city taken a fair library, one hangman (belike fit to execute the fruits of their wits) who had murdered a great number of bodies, would have set fire in it: no, said another very gravely, take heed what you do, for while they are busy about these toys, we shall with more leisure conquer their countries. This indeed is the ordinary doctrine of ignorance, and many words sometimes I have heard spent in it. But because this reason is generally against all learning as well as poetry, or rather, all learning but poetry; because it were too large a digression to handle it, or at least too superfluous (since it is manifest that all government of action is to be gotten by knowledge, and knowledge best by gathering many knowledges, which is reading), I only, with Horace, to him that is of that opinion

jubeo stultum esse libenter;

for as for poetry itself, it is the freest from this objection.

For poetry is the companion of camps. I dare undertake, Orlando Furioso, or honest King Arthur, will never displease a soldier; but the quiddity of *ens* and *prima materia* will hardly agree with a corslet; and therefore, as I said in the beginning, even Turks and Tartars are delighted with poets. Homer, a Greek, flourished before Greece flourished. And if to a slight conjecture a conjecture may be opposed, truly it may seem, that as by him their learned men took almost their first light of knowledge, so their active men received their first motions of courage. Only Alexander's example may serve, who by Plutarch is accounted of such virtue, that

Fortune was not his guide but his footstool; whose acts speak for him, though Plutarch did not: indeed the phoenix of warlike princes. This Alexander left his schoolmaster, living Aristotle, behind him, but took dead Homer with him. He put the philosopher Callisthenes to death for his seeming philosophical, indeed mutinous, stubbornness, but the chief thing he was ever heard to wish for was that Homer had been alive. He well found he received more bravery of mind by the pattern of Achilles than by hearing the definition of fortitude. And therefore, if Cato misliked Fulvius
1140 for carrying Ennius with him to the field, it may be answered that, if Cato misliked it, the noble Fulvius liked it, or else he had not done it; for it was not the excellent Cato Uticensis (whose authority I would much more have reverenced), but it was the former, in truth a bitter punisher of faults (but else a man that had never well sacrificed to the Graces: he misliked and cried out against all Greek learning, and yet, being eighty years old, began to learn it, belike fearing that Pluto understood not Latin). Indeed, the Roman laws allowed no person to be carried to the wars but he that was in the soldier's roll; and therefore, though Cato misliked his unmustered
1150 person, he misliked not his work. And if he had, Scipio Nasica, judged by common consent the best Roman, loved him. Both the other Scipio brothers, who had by their virtues no less surnames than of Asia and Afric, so loved him that they caused his body to be buried in their sepulture. So as Cato's authority, being but against his person, and that answered with so far greater than himself, is herein of no validity.

But now indeed my burden is great; now Plato's name is laid upon me, whom, I must confess, of all philosophers I have ever esteemed most worthy of reverence, and with good reason: since of
1160 all philosophers he is the most poetical. Yet if he will defile the fountain out of which his flowing streams have proceeded, let us boldly examine with what reasons he did it. First, truly, a man might maliciously object that Plato, being a philosopher, was a natural enemy of poets. For indeed, after the philosophers had picked out of the sweet mysteries of poetry the right discerning true points of knowledge, they forthwith putting it in method, and making a school-art of that which the poets did only teach by a divine delightfulness, beginning to spurn at their guides, like ungrateful prentices, were not content to set up shops for themselves, but
1170 sought by all means to discredit their masters; which by the force of delight being barred them, the less they could overthrow them,

the more they hated them. For indeed, they found for Homer seven cities strave who should have him for their citizen; where many cities banished philosophers as not fit members to live among them. For only repeating certain of Euripides' verses, many Athenians had their lives saved of the Syracusans, where the Athenians themselves thought many philosophers unworthy to live. Certain poets, as Simonides and Pindar, had so prevailed with Hiero the First, that of a tyrant they made him a just king; where Plato could do so little with Dionysius, that he himself of a philosopher was made a slave. 1180 But who should do thus, I confess, should requite the objections made against poets with like cavillations against philosophers; as likewise one should do that should bid one read *Phaedrus* or *Symposium* in Plato, or the discourse of love in Plutarch, and see whether any poet do authorize abominable filthiness, as they do. Again, a man might ask out of what commonwealth Plato did banish them: in sooth, thence where he himself alloweth community of women—so as belike this banishment grew not for effeminate wantonness, since little should poetical sonnets be hurtful when a man might have what woman he listed. But I honour philosophical 1190 instructions, and bless the wits which bred them: so as they be not abused, which is likewise stretched to poetry.

St Paul himself (who yet, for the credit of poets, twice citeth poets, and one of them by the name of 'their prophet') setteth a watchword upon philosophy—indeed upon the abuse. So doth Plato upon the abuse, not upon poetry. Plato found fault that the poets of his time filled the world with wrong opinions of the gods, making light tales of that unspotted essence, and therefore would not have the youth depraved with such opinions. Herein may much be said. Let this suffice: the poets did not induce such opinions, 1200 but did imitate those opinions already induced. For all the Greek stories can well testify that the very religion of that time stood upon many and many-fashioned gods, not taught so by the poets, but followed according to their nature of imitation. Who list may read in Plutarch the discourses of Isis and Osiris, of the cause why oracles ceased, of the divine providence, and see whether the theology of that nation stood not upon such dreams which the poets indeed superstitiously observed—and truly (since they had not the light of Christ) did much better in it than the philosophers, who, shaking off superstition, brought in atheism. Plato therefore (whose 1210 authority I had much rather justly construe than unjustly resist) meant not in general of poets, in those words of which Julius

Scaliger saith *Qua authoritate barbari quidam atque hispidi abuti velint ad poetas e republica exigendos*; but only meant to drive out those wrong opinions of the Deity (whereof now, without further law, Christianity hath taken away all the hurtful belief) perchance (as he thought) nourished by the then esteemed poets. And a man need go no further than to Plato himself to know his meaning: who, in his dialogue called *Ion*, giveth high and rightly divine commenda-
1220 tion unto poetry. So as Plato, banishing the abuse, not the thing, not banishing it, but giving due honour unto it, shall be our patron, and not our adversary. For indeed I had much rather (since truly I may do it) show their mistaking of Plato (under whose lion's skin they would make an ass-like braying against poesy) than go about to overthrow his authority; whom, the wiser a man is, the more just cause he shall find to have in admiration; especially since he attributeth unto poesy more than myself do, namely, to be a very inspiring of a divine force, far above man's wit, as in the forenamed dialogue is apparent.

1230 Of the other side, who would show the honours have been by the best sort of judgements granted them, a whole sea of examples would present themselves: Alexanders, Caesars, Scipios, all favourers of poets; Laelius, called the Roman Socrates, himself a poet, so as part of *Heautontimorumenos* in Terence was supposed to be made by him; and even the Greek Socrates, whom Apollo confirmed to be the only wise man, is said to have spent part of his old time in putting Aesop's fables into verses. And therefore, full evil should it become his scholar Plato to put such words in his master's mouth against poets. But what need more? Aristotle writes the Art of
1240 Poesy; and why, if it should not be written? Plutarch teacheth the use to be gathered of them; and how, if they should not be read? And who reads Plutarch's either history or philosophy, shall find he trimmeth both their garments with guards of poesy. But I list not to defend poesy with the help of his underling historiography. Let it suffice to have showed it is a fit soil for praise to dwell upon; and what dispraise may be set upon it, is either easily overcome, or transformed into just commendation.

So that, since the excellencies of it may be so easily and so justly confirmed, and the low-creeping objections so soon trodden down:
1250 it not being an art of lies, but of true doctrine; not of effeminateness, but of notable stirring of courage; not of abusing man's wit, but of strengthening man's wit; not banished, but honoured by Plato: let us rather plant more laurels for to engarland the poets'

heads (which honour of being laureate, whereas besides them only triumphant captains were, is a sufficient authority to show the price they ought to be held in) than suffer the ill-savoured breath of such wrong-speakers once to blow upon the clear springs of poesy.

But since I have run so long a career in this matter, methinks, before I give my pen a full stop, it shall be but a little more lost time to inquire why England, the mother of excellent minds, should be 1260 grown so hard a stepmother to poets, who certainly in wit ought to pass all other, since all only proceedeth from their wit, being indeed makers of themselves, not takers of others. How can I but exclaim

Musa, mihi causas memora, quo numine laeso?

Sweet poesy, that hath anciently had kings, emperors, senators, great captains, such as, besides a thousand others, David, Adrian, Sophocles, Germanicus, not only to favour poets, but to be poets; and of our nearer times can present for her patrons a Robert, king of Sicily, the great King Francis of France, King James of Scotland; such cardinals as Bembus and Bibbiena; such famous preachers 1270 and teachers as Beza and Melanchthon; so learned philosophers as Fracastorius and Scaliger; so great orators as Pontanus and Muretus; so piercing wits as George Buchanan; so grave counsellors as, beside many, but before all, that Hôpital of France, than whom (I think) that realm never brought forth a more accomplished judgement, more firmly builded upon virtue: I say these, with numbers of others, not only to read others' poesies, but to poetize for others' reading—that poesy, thus embraced in all other places, should only find in our time a hard welcome in England, I think the very earth lamenteth it, and therefore decketh our soil with 1280 fewer laurels than it was accustomed. For heretofore poets have in England also flourished, and, which is to be noted, even in those times when the trumpet of Mars did sound loudest. And now that an overfaint quietness should seem to strew the house for poets, they are almost in as good reputation as the mountebanks of Venice. Truly even that, as of the one side it giveth great praise to poesy, which like Venus (but to better purpose) had rather be troubled in the net with Mars than enjoy the homely quiet of Vulcan: so serves it for a piece of a reason why they are less grateful to idle England, which now can scarce endure the pain of a pen. 1290

Upon this necessarily followeth, that base men with servile wits undertake it, who think it enough if they can be rewarded of the printer. And so as Epaminondas is said with the honour of his virtue

to have made an office, by his exercising it, which before was contemptible, to become highly respected; so these men, no more but setting their names to it, by their own disgracefulness disgrace the most graceful poesy. For now, as if all the Muses were got with child to bring forth bastard poets, without any commission they do post over the banks of Helicon, till they make the readers more
1300 weary than post-horses; while, in the meantime, they

Queis meliore luto finxit praecordia Titan

are better content to suppress the outflowings of their wit, than, by publishing them, to be accounted knights of the same order. But I that, before ever I durst aspire unto the dignity, am admitted into the company of the paper-blurrers, do find the very true cause of our wanting estimation is want of desert—taking upon us to be poets in despite of Pallas.

Now, wherein we want desert were a thankworthy labour to express; but if I knew, I should have mended myself. But I, as I
1310 never desired the title, so have I neglected the means to come by it. Only, overmastered by some thoughts, I yielded an inky tribute unto them. Marry, they that delight in poesy itself should seek to know what they do, and how they do; and especially look themselves in an unflattering glass of reason, if they be inclinable unto it. For poesy must not be drawn by the ears; it must be gently led, or rather it must lead—which was partly the cause that made the ancient-learned affirm it was a divine gift, and no human skill: since all other knowledges lie ready for any that hath strength of wit. A poet no industry can make, if his own genius be not carried into it; and
1320 therefore it is an old proverb, *orator fit, poeta nascitur*.

Yet confess I always that as the fertilest ground must be manured, so must the highest-flying wit have a Daedalus to guide him. That Daedalus, they say, both in this and in other, hath three wings to bear itself up into the air of due commendation: that is, art, imitation, and exercise. But these, neither artificial rules nor imitative patterns, we much cumber ourselves withal. Exercise indeed we do, but that very fore-backwardly: for where we should exercise to know, we exercise as having known; and so is our brain delivered of much matter which never was begotten by knowledge.
1330 For there being two principal parts, matter to be expressed by words and words to express the matter, in neither we use art or imitation rightly. Our matter is *quodlibet* indeed, though wrongly performing Ovid's verse,

Quicquid conabor dicere, versus erit;

never marshalling it into any assured rank, that almost the readers cannot tell where to find themselves.

Chaucer, undoubtedly, did excellently in his *Troilus and Criseyde*; of whom, truly, I know not whether to marvel more, either that he in that misty time could see so clearly, or that we in this clear age go so stumblingly after him. Yet had he great wants, fit to be forgiven in so reverent an antiquity. I account the *Mirror of Magistrates* meetly furnished of beautiful parts, and in the Earl of Surrey's lyrics many things tasting of a noble birth, and worthy of a noble mind. The *Shepheardes Calender* hath much poetry in his eclogues, indeed worthy the reading, if I be not deceived. (That same framing of his style to an old rustic language I dare not allow, since neither Theocritus in Greek, Virgil in Latin, nor Sannazaro in Italian did affect it.) Besides these I do not remember to have seen but few (to speak boldly) printed that have poetical sinews in them; for proof whereof, let but most of the verses be put in prose, and then ask the meaning, and it will be found that one verse did but beget another, without ordering at the first what should be at the last; which becomes a confused mass of words, with a tingling sound of rhyme, barely accompanied with reason. 1340 1350

Our tragedies and comedies (not without cause cried out against), observing rules neither of honest civility nor skilful poetry—excepting *Gorboduc* (again, I say, of these that I have seen), which notwithstanding as it is full of stately speeches and well-sounding phrases, climbing to the height of Seneca's style, and as full of notable morality, which it doth most delightfully teach, and so obtain the very end of poesy, yet in truth it is very defectuous in the circumstances, which grieveth me, because it might not remain as an exact model of all tragedies. For it is faulty both in place and time, the two necessary companions of all corporal actions. For where the stage should always represent but one place, and the uttermost time presupposed in it should be, both by Aristotle's precept and common reason, but one day, there is both many days, and many places, inartificially imagined. 1360

But if it be so in *Gorboduc*, how much more in all the rest, where you shall have Asia of the one side, and Afric of the other, and so many other under-kingdoms, that the player, when he cometh in, must ever begin with telling where he is, or else the tale will not be conceived? Now you shall have three ladies walk to gather 1370 flowers: and then we must believe the stage to be a garden. By and by we hear news of shipwreck in the same place: and then we are to blame if we accept it not for a rock. Upon the back of that comes out a hideous monster with fire and smoke: and then the miserable beholders are bound to take it for a cave. While in the meantime two armies fly in, represented with four swords and bucklers: and 1380 then what hard heart will not receive it for a pitched field?

Now, of time they are much more liberal: for ordinary it is that two young princes fall in love; after many traverses, she is got with child, delivered of a fair boy; he is lost, groweth a man, falls in love, and is ready to get another child; and all this in two hours' space: which, how absurd it is in sense, even sense may imagine, and art hath taught, and all ancient examples justified—and at this day, the ordinary players in Italy will not err in. Yet will some bring in an example of *Eunuchus* in Terence, that containeth matter of two days, yet far short of twenty years. True it is, and so was it to be played 1390 in two days, and so fitted to the time it set forth. And though Plautus have in one place done amiss, let us hit with him, and not miss with him.

But they will say: How then shall we set forth a story which containeth both many places and many times? And do they not know that a tragedy is tied to the laws of poesy, and not of history; not bound to follow the story, but having liberty either to feign a quite new matter or to frame the history to the most tragical conveniency? Again, many things may be told which cannot be showed, if they know the difference betwixt reporting and repre- 1400 senting. As, for example, I may speak (though I am here) of Peru, and in speech digress from that to the description of Calicut; but in action I cannot represent it without Pacolet's horse; and so was the manner the ancients took, by some *Nuntius* to recount things done in former time or other place. Lastly, if they will represent a history, they must not (as Horace saith) begin *ab ovo*, but they must come to the principal point of that one action which they will represent.

By example this will be best expressed. I have a story of young Polydorus, delivered for safety's sake, with great riches, by his 1410 father Priam to Polymnestor, king of Thrace, in the Trojan war time; he, after some years, hearing the overthrow of Priam, for to make the treasure his own, murdereth the child; the body of the child is taken up by Hecuba; she, the same day, findeth a sleight to be revenged most cruelly of the tyrant. Where now would one of our tragedy writers begin, but with the delivery of the child? Then should he sail over into Thrace, and so spend I know not how many years, and travel numbers of places. But where doth Euripides? Even with the finding of the body, leaving the rest to be told by the spirit of Polydorus. This need no further to be enlarged; the dullest wit may conceive it. 1420

But besides these gross absurdities, how all their plays be neither right tragedies, nor right comedies, mingling kings and clowns, not because the matter so carrieth it, but thrust in the clown by head and shoulders to play a part in majestical matters with neither decency nor discretion, so as neither the admiration and commiseration, nor the right sportfulness, is by their mongrel tragicomedy obtained. I know Apuleius did somewhat so, but that is a thing recounted with space of time, not represented in one moment; and I know the ancients have one or two examples of tragicomedies, as Plautus hath *Amphitryo*; but, if we mark them well, we 1430 shall find that they never, or very daintily, match hornpipes and funerals. So falleth it out that, having indeed no right comedy, in that comical part of our tragedy, we have nothing but scurrility, unworthy of any chaste ears, or some extreme show of doltishness, indeed fit to lift up a loud laughter, and nothing else: where the whole tract of a comedy should be full of delight, as the tragedy should be still maintained in a well-raised admiration.

But our comedians think there is no delight without laughter; which is very wrong, for though laughter may come with delight, yet cometh it not of delight, as though delight should be the cause 1440 of laughter; but well may one thing breed both together. Nay, rather in themselves they have, as it were, a kind of contrariety: for delight we scarcely do but in things that have a conveniency to ourselves or to the general nature; laughter almost ever cometh of things most disproportioned to ourselves and nature. Delight hath a joy in it, either permanent or present. Laughter hath only a scornful tickling.

For example, we are ravished with delight to see a fair woman, and yet are far from being moved to laughter; we laugh at deformed creatures, wherein certainly we cannot delight. We delight in good 1450 chances, we laugh at mischances: we delight to hear the happiness of our friends, or country, at which he were worthy to be laughed at that would laugh; we shall, contrarily, laugh sometimes to find a matter quite mistaken and go down the hill against the bias in the mouth of some such men—as for the respect of them one shall be heartily sorry, he cannot choose but laugh, and so is rather pained than delighted with laughter.

Yet deny I not but that they may go well together. For as in Alexander's picture well set out we delight without laughter, and in 1460 twenty mad antics we laugh without delight; so in Hercules, painted with his great beard and furious countenance, in a woman's attire, spinning at Omphale's commandment, it breedeth both delight and laughter: for the representing of so strange a power in love procureth delight, and the scornfulness of the action stirreth laughter. But I speak to this purpose that all the end of the comical part be not upon such scornful matters as stir laughter only, but, mixed with it, that delightful teaching which is the end of poesy. And the great fault even in that point of laughter, and forbidden plainly by Aristotle, is that they stir laughter in sinful things, which 1470 are rather execrable than ridiculous, or in miserable, which are rather to be pitied than scorned. For what is it to make folks gape at a wretched beggar and a beggarly clown; or, against law of hospitality, to jest at strangers, because they speak not English so well as we do? What do we learn, since it is certain

> Nil habet infelix paupertas durius in se,
> Quam quod ridiculos homines facit?

But rather, a busy loving courtier and a heartless threatening Thraso; a self-wise-seeming schoolmaster; an awry-transformed traveller. These, if we saw walk in stage names, which we play 1480 naturally, therein were delightful laughter, and teaching delightfulness—as in the other, the tragedies of Buchanan do justly bring forth a divine admiration.

But I have lavished out too many words of this play matter. I do it because, as they are excelling parts of poesy, so is there none so much used in England, and none can be more pitifully abused; which, like an unmannerly daughter showing a bad education, causeth her mother Poesy's honesty to be called in question.

Other sort of poetry almost have we none, but that lyrical kind of songs and sonnets: which, Lord, if He gave us so good minds, 1490 how well it might be employed, and with how heavenly fruit, both private and public, in singing the praises of the immortal beauty: the immortal goodness of that God who giveth us hands to write and wits to conceive; of which we might well want words, but never matter; of which we could turn our eyes to nothing, but we should ever have new-budding occasions. But truly many of such writings as come under the banner of unresistible love, if I were a mistress, would never persuade me they were in love: so coldly they apply fiery speeches, as men that had rather read lovers' writings—and so caught up certain swelling phrases which hang together, like a man that once told my father that the wind was at north-west and by south, because he would be sure to name winds enough—than that in truth they feel those passions, which easily (as I think) may be bewrayed by that same forcibleness or *energia* (as the Greeks call it) of the writer. But let this be a sufficient though short note, that we miss the right use of the material point of poesy. 1500

Now, for the outside of it, which is words, or (as I may term it) diction, it is even well worse. So is that honey-flowing matron Eloquence apparelled, or rather disguised, in a courtesan-like painted affectation: one time, with so far-fet words that may seem monsters but must seem strangers to any poor Englishman; another time, with coursing of a letter, as if they were bound to follow the method of a dictionary; another time, with figures and flowers, extremely winter-starved. But I would this fault were only peculiar to versifiers, and had not as large possession among prose-printers; and (which is to be marvelled) among many scholars; and (which is to be pitied) among some preachers. Truly I could wish, if at least I might be so bold to wish a thing beyond the reach of my capacity, the diligent imitators of Tully and Demosthenes (most worthy to be imitated) did not so much keep Nizolian paper-books of their figures and phrases, as by attentive translation (as it were) devour them whole, and make them wholly theirs: for now they cast sugar and spice upon every dish that is served to the table—like those Indians, not content to wear earrings at the fit and natural place of the ears, but they will thrust jewels through their nose and lips, because they will be sure to be fine. Tully, when he was to drive out Catiline, as it were with a thunderbolt of eloquence, often used the figure of repetition, as *Vivit. Vivit? Imo in senatum venit, &c.* Indeed, inflamed with a well-grounded rage, he would have his words (as it were) double out of his mouth, and so do that artificially which we see men in choler do naturally. And we, having noted the grace of those words, hale them in sometimes to a familiar epistle, when it were too too much choler to be choleric. How well store of *similiter cadences* doth sound with the gravity of the pulpit, I would but invoke Demosthenes' soul to tell, who with a rare daintiness useth them. Truly they have made me think of the sophister that with too much subtlety would prove two eggs three, and though he 1510 1520 1530

might be counted a sophister, had none for his labour. So these
men bringing in such a kind of eloquence, well may they obtain an
opinion of a seeming finesse, but persuade few—which should be
1540 the end of their finesse. Now for similitudes, in certain printed
discourses, I think all herbarists, all stories of beasts, fowls, and
fishes are rifled up, that they come in multitudes to wait upon any
of our conceits; which certainly is as absurd a surfeit to the ears as
is possible. For the force of a similitude not being to prove anything
to a contrary disputer, but only to explain to a willing hearer, when
that is done, the rest is a most tedious prattling, rather over-swaying
the memory from the purpose whereto they were applied, than any
whit informing the judgement, already either satisfied, or by
similitudes not to be satisfied. For my part, I do no doubt, when
1550 Antonius and Crassus, the great forefathers of Cicero in eloquence,
the one (as Cicero testifieth of them) pretended not to know art,
the other not to set by it, because with a plain sensibleness they
might win credit of popular ears (which credit is the nearest step
to persuasion, which persuasion is the chief mark of oratory), I do
not doubt (I say) but that they used these knacks very sparingly;
which who doth generally use, any man may see doth dance to his
own music, and so be noted by the audience more careful to speak
curiously than to speak truly. Undoubtedly (at least to my opinion
undoubtedly), I have found in divers smally learned courtiers a
1560 more sound style than in some professors of learning; of which I
can guess no other cause, but that the courtier, following that which
by practice he findeth fittest to nature, therein (though he know it
not) doth according to art, though not by art: where the other, using
art to show art, and not to hide art (as in these cases he should do),
flieth from nature, and indeed abuseth art.

But what? Methinks I deserve to be pounded for straying from
poetry to oratory. But both have such an affinity in the wordish
consideration, that I think this digression will make my meaning
receive the fuller understanding: which is not to take upon me to
1570 teach poets how they should do, but only, finding myself sick among
the rest, to show some one or two spots of the common infection
grown among the most part of writers, that, acknowledging ourselves somewhat awry, we may bend to the right use both of matter
and manner: whereto our language giveth us great occasion, being
indeed capable of any excellent exercising of it. I know some will
say it is a mingled language. And why not so much the better, taking
the best of both the other? Another will say it wanteth grammar.

Nay truly, it hath that praise, that it wants not grammar: for grammar it might have, but it needs it not, being so easy in itself, and so void of those cumbersome differences of cases, genders, 1580 moods, and tenses, which I think was a piece of the Tower of Babylon's curse, that a man should be put to school to learn his mother-tongue. But for the uttering sweetly and properly the conceits of the mind (which is the end of speech), that hath it equally with any other tongue in the world; and is particularly happy in compositions of two or three words together, near the Greek, far beyond the Latin, which is one of the greatest beauties can be in a language.

Now of versifying there are two sorts, the one ancient, the other modern: the ancient marked the quantity of each syllable, and 1590 according to that framed his verse; the modern, observing only number (with some regard of the accent), the chief life of it standeth in that like sounding of the words, which we call rhyme. Whether of these be the more excellent, would bear many speeches: the ancient (no doubt) more fit for music, both words and time observing quantity, and more fit lively to express diverse passions, by the low or lofty sound of the well-weighed syllable; the latter likewise, with his rhyme, striketh a certain music to the ear, and, in fine, since it doth delight, though by another way, it obtains the same purpose: there being in either sweetness, and wanting in 1600 neither majesty. Truly the English, before any vulgar language I know, is fit for both sorts. For, for the ancient, the Italian is so full of vowels that it must ever be cumbered with elisions; the Dutch so, of the other side, with consonants, that they cannot yield the sweet sliding, fit for a verse; the French in his whole language hath not one word that hath his accent in the last syllable saving two, called *antepenultima*; and little more hath the Spanish, and therefore very gracelessly may they use dactyls. The English is subject to none of these defects. Now for the rhyme, though we do not observe quantity, yet we observe the accent very precisely, which 1610 other languages either cannot do, or will not do so absolutely. That *caesura*, or breathing place in the midst of the verse, neither Italian nor Spanish have, the French and we never almost fail of. Lastly, even the very rhyme itself, the Italian cannot put it in the last syllable, by the French named the masculine rhyme, but still in the next to the last, which the French call the female, or the next before that, which the Italian term *sdrucciola*. The example of the former is *buono: suono*, of the *sdrucciola* is *femina: semina*. The French, of

the other side, hath both the male, as *bon: son*, and the female, as
1620 *plaise: taise*, but the *sdrucciola* he hath not: where the English hath all three, as *due: true, father: rather, motion: potion*—with much more which might be said, but that already I find the triflingness of this discourse is much too much enlarged.

So that since the ever-praiseworthy Poesy is full of virtue-breeding delightfulness, and void of no gift that ought to be in the noble name of learning; since the blames laid against it are either false or feeble; since the cause why it is not esteemed in England is the fault of poet-apes, not poets; since, lastly, our tongue is most fit to honour poesy, and to be honoured by poesy; I conjure you all that
1630 have had the evil luck to read this ink-wasting toy of mine, even in the name of the nine Muses, no more to scorn the sacred mysteries of poesy; no more to laugh at the name of poets, as though they were next inheritors to fools; no more to jest at the reverent title of a rhymer; but to believe, with Aristotle, that they were the ancient treasurers of the Grecians' divinity; to believe, with Bembus, that they were first bringers-in of all civility; to believe, with Scaliger, that no philosopher's precepts can sooner make you an honest man than the reading of Virgil; to believe, with Clauserus, the translator of Cornutus, that it pleased the heavenly Deity, by Hesiod and
1640 Homer, under the veil of fables, to give us all knowledge, logic, rhetoric, philosophy natural and moral, and *quid non?*; to believe, with me, that there are many mysteries contained in poetry, which of purpose were written darkly, lest by profane wits it should be abused; to believe, with Landino, that they are so beloved of the gods that whatsoever they write proceeds of a divine fury; lastly, to believe themselves, when they tell you they will make you immortal by their verses. Thus doing, your name shall flourish in the printers' shop; thus doing, you shall be of kin to many a poetical preface; thus doing, you shall be most fair, most rich, most wise, most all,
1650 you shall dwell upon superlatives; thus doing, though you be *libertino patre natus*, you shall suddenly grow *Herculea proles*,

Si quid mea carmina possunt;

thus doing, your soul shall be placed with Dante's Beatrice, or Virgil's Anchises. But if (fie of such a but) you be born so near the dull-making cataract of Nilus that you cannot hear the planet-like music of poetry; if you have so earth-creeping a mind that it cannot lift itself up to look to the sky of poetry, or rather, by a certain rustical disdain, will become such a mome as to be a Momus of poetry; then, though I will not wish unto you the ass's ears of Midas, nor to be driven by a poet's verses, as Bubonax was, to hang himself, 1660 nor to be rhymed to death, as is said to be done in Ireland; yet thus much curse I must send you, in the behalf of all poets, that while you live, you live in love, and never get favour for lacking skill of a sonnet; and, when you die, your memory die from the earth for want of an epitaph.

Notes

ABBREVIATIONS

Place of publication is London unless otherwise specified.

Aesop, *Fables*	J. R. Turner (ed.), *The Works of William Bullokar*, vol. iv, *Aesop's Fablz 1585* (1969)
AS	*Astrophil and Stella*
Buxton	John Buxton, *Sir Philip Sidney and the English Renaissance* (1964)
CS	*Certain Sonnets*
Diana	Judith M. Kennedy, *A Critical Edition of Yong's translation of George of Montemayor's Diana and Gil Polo's Enamoured Diana* (1968)
DNB	*Dictionary of National Biography*
DP	*The Defence of Poesy*
Dyer	Ralph M. Sargent, *At the Court of Queen Elizabeth: The Life and Lyrics of Sir Edward Dyer* (1935)
ELH	*English Literary History*
ELR	*English Literary Renaissance*
Feuillerat	Albert Feuillerat (ed.), *The Works of Sir Philip Sidney* (4 vols., 1912–26); repr. as *Prose Works* (1962)
Fraunce	Abraham Fraunce, *The Arcadian Rhetorike [1588]*, ed. Ethel Seaton, Luttrell Society Reprints 9 (1950)
Greville, *Prose Works*	John Gouws (ed.), *The Prose Works of Fulke Greville, Lord Brooke* (1986)
Historical Remembrance	Edmund Molyneux, *Historical Remembrance of the Sidneys, the father and the son*, from R. Holinshed, *The third volume of Chronicles* (1588)
HMC	Calendars of manuscripts published by the Historical Manuscripts Commission
JWCI	*Journal of the Warburg and Courtauld Institutes*
Kay	Dennis Kay (ed.), *Sir Philip Sidney: An Anthology of Modern Criticism* (1987)

'Lady Rich'	Katherine Duncan-Jones, 'Sidney, Stella and Lady Rich', in J. van Dorsten, D. Baker Smith, and A. F. Kinney (eds.), *Sir Philip Sidney: 1586 and the Creation of a Legend* (Leiden, 1986)
Lant, *Roll*	Thomas Lant, *Sequitur celebritas et pompa funebris* (1587), reproduced in A. M. Hind, *Engraving in England in the Sixteenth and Seventeenth Centuries* (1952), vol. i.
LM	*The Lady of May*
Misc. Prose	K. Duncan-Jones and J. van Dorsten (eds.), *Miscellaneous Prose of Sir Philip Sidney* (1973)
Moffet, *Nobilis*	V. B. Heltzel and H. H. Hudson (eds.), *Thomas Moffet: Nobilis, Or, A View of the Life and Death of a Sidney* (San Marino, Calif., 1940)
NA	V. J. Skretkowicz (ed.), *The Countess of Pembroke's Arcadia (The New Arcadia)* (1987)
Nichols, *Progresses*	John Nichols (ed.), *The Progresses and Public Processions of Queen Elizabeth* (3 vols., 1823)
North's *Plutarch*	W. E. Henley (ed.), *Plutarch's Lives of the Noble Grecians and Romans Englished by Sir Thomas North anno 1579, Tudor Translations* vii (6 vols., 1895)
OA	Jean Roberston (ed.), *Sir Philip Sidney: The Countess of Pembroke's Arcadia (The Old Arcadia)* (1973)
OED	*The Oxford English Dictionary*
OP	*Other Poems* (Ringler)
Osborn	James M. Osborn, *Young Philip Sidney 1572–1577* (New Haven, Conn., and London, 1972)
Ottley	Peter Beal, 'Poems by Sir Philip Sidney. The Ottley Manuscript', *The Library*, fifth series, 33 (1978), 284–95
PQ	*Philological Quarterly*
RES	*Review of English Studies*
Ringler	W. A. Ringler (ed.), *The Poems of Sir Philip Sidney* (1962)
Robert Sidney	Sir Robert Sidney, *Poems*, ed. P. J. Croft (1984)
Sannazaro	Jacopo Sannazaro, *Arcadia* [1504], translated by R. Nash (Detroit, 1966)
Shepherd	Geoffrey Shepherd (ed.), *Sidney: An Apology for Poetry* (1965)

Temple, *Analysis*	John Webster (ed.), '*William Temple's Analysis of Sir Philip Sidney's Apology for Poetry* (New York 1984)
Tilley	Morris Palmer Tilley, *A Dictionary of Proverbs in England in the Sixteenth and Seventeenth Centuries* (Ann Arbor, Mich., 1950)
Triumph	Henry Goldwell, *A brief declaration of the shows performed before the Queen's Majesty and the French ambassadors* (1581) (also called *The Triumph of the Four Foster Children of Desire*)
van Dorsten	J. van Dorsten, D. Baker Smith, and A. F. Kinney (eds.), *Sir Philip Sidney: 1586 and the Creation of a Legend* (Leiden, 1986)
Wallace	Malcolm W. Wallace, *The Life of Sir Philip Sidney* (1915)
Young	R. B. Young, 'English Petrarke' in *Three Studies in the Renaissance* (New Haven, Conn., 1958)

1 *The Lady of May*, untitled in early texts, was included as the final item in all the folio editions of the *Arcadia* from 1599 onwards. There is also a manuscript text, formerly at Helmingham Hall, Norfolk, now BL Add. MS 61821, which is the sole source for Rombus's final speech. The setting of the entertainment, Wanstead Manor in Essex, had been bought from Lord Rich by Sidney's uncle, the Earl of Leicester, in 1577. The Queen visited him there in May 1578 and May 1579, and it is impossible to determine to which year *LM* belongs. If it was performed in the latter year it must surely have some bearing on Leicester's marriage to the widowed Countess of Essex, Lettice Knollys, of which the Queen had learned in January, and possibly also, as Stephen Orgel and others have argued, on the Queen's own courtship by the Duke of Alençon.

ll. 2–3. *one apparelled like an honest man's wife.* Leicester is described by Rombus, l. 346 below, as 'an honest man'; if *LM* belongs to 1579 it is possible that it was designed by Sidney as a vehicle for the presentation to the Queen of Leicester's new wife and her eldest daughter, Penelope Devereux (later to become Lady Rich). 'One apparelled like' may hint that the woman suitor's true identity is more exalted than her rustic dress suggests.

l. 4. *desiring all the lords and gentlemen to speak a good word for her.* The new Countess of Leicester was much in need of such support.

l. 17. *honesty.* Chastity.

l. 21. *partakers.* Supporters, seconds.

1 l. 26. *infectious.* Destructive.

Supplication. The Supplication is really contained in the preceding speech, rather than in this verse address.

2 l. 45. *fosters.* Foresters.

l. 47. *Master Rombus.* From *rhombus*, a figure whose four sides and opposite angles are equal. Fielding used the same idea of angularity and pedantry in calling one of Tom Jones's tutors 'Square'.

l. 51. *startle.* Start with surprise or alarm.

l. 52. *old father Lalus.* In the Third Eclogues of *OA* the marriage of young Lalus to Kala is celebrated; perhaps his father is to be imagined as having the same name.

l. 58. *minsical.* Mincing.

l. 60. *featioust.* Best formed, most handsome; cf. Latin *facticius.*

l. 61. *By my mother Kit's soul.* An invented oath.

l. 61. *fransical.* Frenzied. Sidney preferred the spelling 'franzy'; cf. *CS* 30.

l. 64. *loquence.* Fluency of speech.

l. 66. *bashless.* Bold, shameless, but used here in the reverse sense.

l. 71. *transfund his dotes.* Pour out his gifts, perhaps also with some play on Jove's thundering 'darts'.

3 l. 76. *juvental fry.* Young infants.

l. 77. *mansuetude.* Gentleness.

l. 79. *Parcere subjectis et debellare superbos.* Virgil, *Aeneid*, vi. 853, 'to spare the humble and cast down the proud'.

l. 81. *solummodo.* Alone.

l. 82. *sanguinolent.* Bloody.

l. 83. *Pecorius Asinus.* For *pecus asininus,* asinine brute.

l. 84. *Dixi. Verbum sapiento satum est.* 'I have spoken; a word to the wise is enough' (Tilley W781).

ll. 85–6. *sulks of the sandiferous seas.* Furrows of the sand-bearing seas.

l. 86. *Haec olim meminisse iuvabit.* Virgil, *Aeneid,* I. 203, 'one day we shall enjoy the recollection of these things'.

ll. 86–7. *ad propositos revertebo.* 'I shall return to the matter in hand.'

l. 90. *quodammodo.* In some manner.

l. 91. *a cast.* Technical term from hawking for the number of hawks cast off at one time, i.e. two.

3 l. 99. *O Tempori, O Moribus.* Garbled from '*O tempora, o mores*', 'O times! O manners!' (Cicero, *In Catilinam*, I.i.)

l. 100. *turpify.* Befoul.

4 ll. 110–11. *a certain gentleman hereby.* Leicester.

l. 118. *that wherein.* In beauty.

l. 125. *With me have been.* The parenthesis suggests a mild innuendo.

l. 126. *Therion.* The name means 'wild creature'.

l. 126. *Espilus.* 'Felt-presser'; hence, one who handles wool.

l. 152. *baldrics.* Belts worn across the shoulder.

l. 172. *Them can I take, but you I cannot hold.* Therion uses the traditional comparison of an elusive lady to a deer, frequent in medieval poetry, and no doubt familiar to Sidney from Petrarch, Wyatt, and others; cf. 'and wild for to hold, though I seem tame' (Wyatt, *Poems*, ed. R. A. Rebholz (1978), 77).

6 ll. 183–4. *Dorcas . . . Rixus.* 'Gazelle' (i.e. gentle) and 'quarrelsome'.

l. 189. *the harlotry.* The harlot.

l. 190. *the sheep's rot.* Liver disease still known to farmers. The implied identification of the lady with a sheep parallels Therion's identification of her with a deer, above.

l. 192. *O Midas.* The idea is presumably that the ass's ears of Midas would be appropriate for listening to drivel.

l. 194. *blaying.* Bleating; cf. *AS* 9. 49–50.

ll. 212–13. *Heu, Ehem, Hei, Insipidum, Inscitium vulgorum et populorum.* A Latin tag may underlie this semi-gibberish—'Woe on you, ignorant rabble'—but it has not been traced.

l. 213. *nebulons.* From *nebulo*, paltry, worthless fellow.

l. 214. *edify.* Establish.

l. 215. *throw your ears to me.* A literal rendering of Latin idioms such as *praebere aures, admovere aures,* 'lend me your ears'.

l. 216. *indoctrinated your plumbeous cerebrosities.* Instructed your leaden brains.

6–7 ll. 219–20. *prius . . . gratia.* 'First a speech must be divided, before it is defined; for instance'; '*exemplum gratia*' is an error for *exempli gratia*, or 'e.g.'.

7 l. 241. *templer.* Probably intended to suggest 'one who contemplates', as well, perhaps, as 'knight templar', suggesting religious zeal, or 'barrister', or inhabitant of the Inner or Middle Temple, suggesting diligent study.

l. 258. *dilucidate.* Elucidate; for once, a correct usage.

8 l. 261. *equitate*. This should mean 'ride, as on a horse'; presumably Rombus means 'make a confusing equation'.

ll. 262–3. *an enthymeme a loco contingentibus*. An argument from proximity of position.

l. 265. *Darius King of Persia*. A reference to the mnemonic word 'Darii' in formal logic, denoting the mood 'in which the major premiss is a universal affirmative (*a*), and the minor premiss and the conclusion particular affirmatives (*i*)' (Thomas Wilson, *The Rule of Reason* (1567), f. 27). The first four 'modes' or 'moods' are denoted by the words *barbara, celarent, Darii, ferio*.

l. 268. *his major . . . is a fool*. Terms from logic, for the major and minor premiss in a syllogism.

l. 269. *et ecce homo blancatus quasi lilium*. 'And behold a man lily-white!' Perhaps a quotation, but not traced. The implication may be that Dorcas will turn white with dismay.

9 l. 311. *Sylvanus*. The god of foresters. George Gascoigne appeared as Sylvanus in Leicester's Kenilworth entertainment, at which Sidney was present, in 1575 (Nichols, *Progresses*, ii. 515).

l. 317. *Pan*. The god of shepherds. Sidney used the story of Pan mistaking the bed of Hercules for that of his mistress Omphale in *OA* 225.

10 l. 338. *round agates*. Agates were commonly carved, sometimes into human figures, as references in Shakespeare show (2 *Henry IV*, I. ii. 90; *Much Ado*, III. i. 65). But these were presumably smoothly rounded. Agates are only semi-precious, and Sidney and Leicester may have exploited the rustic fiction of *LM* to give the Queen a cheaper present than would normally have been acceptable.

l. 340. *barbarons*. Barbarians, perhaps on the (false) analogy of nebulons, above, 213.

l. 341. *vapilated*. Whipped.

l. 343. *Juno, Venus, Pallas et profecto plus*. A reference to the Judgement of Paris, implying that the Queen has the virtue of all three goddesses and more as well. Cf. the picture of Queen Elizabeth at Hampton Court (by Hans Eworth, *c*. 1569) in which Elizabeth, confronting the three goddesses, awards the apple to herself; also George Peele, *The Arraignment of Paris* (1584).

l. 347. *ædicle*. From Latin *aediculum*, small house.

l. 348. *oves, boves et pecora campi*. From the Vulgate, Ps. 8. 8. The Prayer Book version is 'All sheep and oxen: yea, and the beasts of the field'.

l. 350. *O heu Aedipus Aecastor*. Perhaps a garbled version of some exclamation in Seneca's *Oedipus* (translated by Alexander Neville in

1563), or *Thebais* (translated by Newton in 1581), or even Gascoigne's *Jocasta* (1575). The uncertainties of this part of the text, based on a single manuscript version, are too great to warrant emendation to 'Oedipus, Jocasta'.

10 l. 355. *saith 'and Elizabeth'.* The suggestion may be either that Leicester concludes the Lord's Prayer 'In the name of the Father, the Son, the Holy Ghost and Elizabeth', or that he substitutes a prayer to Elizabeth for *Ave Maria.*

l. 356. *secundum the civil law.* According to the civil law. The three categories are natural law, the law of nations, and civil law.

ll. 357–8. *nine hundredth paragroper of the 7.ii. code in the great Turk Justinian's library.* A reference to the *Codex Justinian*; 'code' was the usual term for the collections of statutes of Justinian or Theodosius (*OED*). Lib. vii., tit. ii is *De testamentaria manumissione*, which seems appropriate; Rombus is not always as muddled as he seems. 'Nine hundredth paragroper' is probably an invented reference intended to increase the impressiveness of the citation.

l. 358. *deponed all his juriousdiction.* Relinquished all his rights.

l. 359. *tibi dominorum domina.* To you, mistress of masters.

ll. 360–1. *iure gentiorum.* By the law of nations.

11 *Song: 'Sleep, baby mine, desire'.* The only poem of Sidney's to survive in his own hand, inscribed on the last leaf of a copy of Jean Bouchet, *Les Annales d'Aquitaine* (Poitiers, 1557), now in the Biblioteca Bodmeriana, Cologny, Geneva (Reproduced and discussed in P. J. Croft, *Autograph Poetry in the English Language* (1973), i. 14–15). Sidney's inscription of the lyric in a book apparently belonging to someone else is added evidence of the 'social' quality of *CS* (see above). There are many settings of the Italian song '*Basciami vita mia*' from 1543 onwards, and it is impossible to know which Sidney had in mind.

l. 3. *Way.* Go away; but also an onomatopoeic suggestion of the baby's cry.

l. 9. *for that.* For food, i.e. satisfaction of desire.

Song: 'Who hath his fancy pleased'. The tune to which the poem is set was originally that of a French Catholic song composed in mockery of the Huguenot Prince Louis de Condé in 1568; it was adopted by the Prince de Condé's own troops, became the song of the House of Orange, and is now that of the Dutch National Anthem. Sidney 'may have heard the song during his visit to the Prince of Orange in 1577' (Ringler, 431). The theme, unusually for Sidney, seems to be Neo-Platonic.

12 *'Ring out your bells'*. One of Sidney's most popular poems, which was already in manuscript circulation during his lifetime (Ringler, 555) and was included in *England's Helicon* (1600). Its enduring popularity is shown by Tennyson's protest at source-hunters who 'will not allow one to say "Ring the bell" without finding that we have taken it from Sir P. Sidney' (Hallam Tennyson, *Alfred Lord Tennyson: A Memoir* [1897]. i. 258).

l. 1. *mourning shows*: black cloths.

l. 7. *ungrateful:* cf. *AS* 31.14.

13 l. 21. *trentals*. A sequence of thirty requiem masses.

l. 33. *Love is not dead, but sleepeth*. Cf. 'the maid is not dead, but sleepeth' (Matt. 9: 24).

Sonnet: 'Thou blind man's mark'

l. 3. *Band*. Swaddling band.

l. 4. *web*. 'Cloth in the process of being woven, suggesting Penelope's web' (Ringler, 434).

Sonnet: 'Leave me, O love'

14 l. 6. *sweet yoke*. Image which identifies the alternative to love which 'reachest but to dust' as divine, since it refers to Christ's words 'my yoke is easy, and my burden light' (Matt. 11: 30). Sidney also knew an *impresa* consisting of a picture of a yoke with the motto *Sauve* (Sweet), used by Pope Leo X (Paolo Giovio, *Dialogo dell'Imprese* (Lyons, 1574) 45; Bodleian MS Rawl, D. 345, fo. 28).

l. 15. *Splendidis longum valedico nugis*. The source of this motto, 'I bid a long farewell to splendid trifles', has not been traced. In Ottley (292) *CS* 32 is followed by a motto known to have been used elsewhere by Sidney, *Virtus secura sequatur*, 'let virtue follow in safety' (K. Duncan-Jones, 'Sidney's Personal *Imprese*', *JWCI* 33 (1970), 323).

15 *The lad Philisides*. Probably written in 1577–80, when Sidney wrote other poems about Philisides and Mira. It appears to be located somewhere in Eastern England, reached by wind from 'Holland' (106). Wanstead, at Essex, the setting for *LM*, is to the east of London. It was first printed in the 1593 *Arcadia*, inserted, somewhat inappropriately, in the Third Eclogues. It is the first *canzone* to be written in English, being based, rather loosely, on Ecloga 3 of Sannazaro's *Arcadia*. Whereas Sannazaro's *canzone* is sung on 3 Mar., Sidney's appears to belong to high spring, and might therefore belong to May celebrations at Wanstead in 1578 or 1579.

16 l. 42. *make*. Mate.

16 l. 45. *engender children high.* Probably referring to the idea that bees were born from the dew on flowers.

17 l. 78. *bea-waymenting.* Lamenting with bleats.

18 l. 113. *willow's bark.* Willow was worn by forsaken lovers; cf. *OA* 13.118.

l. 125. *westward eyeing.* Mira is to the west of Philisides (see above); but this does not really tell us much about her identity or location.

19 *'What tongue can her perfection tell'.* This poem comes near the end of Book Three of *OA*, where it is described as 'a song the shepherd Philisides had . . . sung of the beauties of his unkind mistress'. The lyric comes into the mind of Pyrocles as he lays Philoclea on her bed; while the audience of 'fair ladies' enjoy this poetic celebration of Philoclea's body, Pyrocles enjoys the real thing. *OA* MSS show that Sidney worked hard over this light, easy-seeming Ovidian *blason*, or catalogue of beauties, revising it several times. It quickly became one of his most popular poems, occurring in many MS commonplace books and, for instance, in *England's Parnassus* (1600). Puttenham described it as 'excellently well handled' (*Arte of English Poesy*, ed. G. D. Willcock and A. Walker (1936), 244), and many later poets imitated it; see, for instance, Lord Herbert of Cherbury, *Poems*, ed. G. C. Moore Smith (1923), 2–5; Thomas Carew, *Poems*, ed. Rhodes Dunlap (1949), 99–101.

l. 16. *Their matchless praise.* This is the reading of *OA* MSS. Robertson follows the 1590 edition in giving 'The matchless pair'; however, this may be a case of early editorial emendation. 'Matchless praise' fits in with the next four lines, in which no simile of 'praise' is adequate to 'praise' her eyes; cf. also *AS* 35.13–14.

l. 25. *queen-apple.* A kind of red apple.

20 l. 51. *a say.* An assay, a foretaste.

l. 57. *porphyry.* A reddish crystal.

21 l. 97. *bought incaved.* Inbent curve.

l. 98. *like cunning painter.* Perhaps a reference to miniature painting, in which density was given to colour by the use of white.

l. 99. *the gart'ring place.* Just above the knee, where garters were tied.

l. 103. *Atlas.* The ankle, which supports her heavenly body, as Atlas was supposed to support the heavens.

l. 110. *mews.* Moults.

l. 116. *the hate-spot ermelin.* The white ermine was thought to hate dirt so much that it would die rather than allow its coat to be stained.

There is an *impresa* of an ermine with the motto *'Rather dead than spotted'* in *NA* (101).

22 *Epithalamium: Let mother earth.* One of the earliest epithalamia in English. The verse form is based on that of a wedding poem in *Diana* (378–9). We should notice that Cupid, who dominates the extra-marital loves of the princes and the Arcadian royal family, is here firmly banished, in favour of Hymen. This is consistent with Dicus's bitter attack on Cupid in the First Eclogues (49–50).

Philisides' fable: 'As I my little flock'

25 l. 1. *As I my little flock on Ister bank.* This poem reflects Sidney's friendship with the Huguenot statesman Hubert Languet, who instructed him in political wisdom. Sidney and Languet were together on 'Ister bank', i.e. in Vienna, on the Danube, in Aug. 1573 and Aug. 1574, but there is no way of knowing either which year the poem refers to or when it was written. It is the poem which brings him closest to Spenser, being written in 'old rustic language', such as he was to criticize in *DP* (l. 1346). Specifically, it is analogous to the February Eclogue of *The Shepheardes Calender* (1579), a beast fable taught to the poet Thenot in his youth by an older poet, Tityrus.

l. 2. *couthe.* Knew.

26 l. 24. *naught.* Evil.

l. 29. *thilk.* Those.

l. 30. *jump.* Exact, perfect.

l. 33. *old true tales.* Writing to Sidney in 1579, Daniel Rogers referred to Languet as he who 'guided you through the histories and origins of states' (*OA* 463).

l. 40. *worthy Coredens.* Probably Edward Dyer (1543–1607).

l. 43. *Such manner time there was.* This passage derives partly from Ovid's accounts of the Golden Age (*Metamorphoses,* i and xv) and partly from Isa. II: 6–8. Greville recalls Sidney's myth in 'A Treatise of Monarchy', stanza 122 (*Remains*, ed. G. Wilkes (1965), 65). *I not.* I know not.

l. 45. *woned.* Inhabited.

27 l. 51. *nis.* Is not.

l. 58. *envy harb'reth most in feeble hearts.* A parody of the Chaucerian commonplace, 'Pittee renneth sone in gentil herte'.

l. 66. *pewing.* Plaintive crying.

l. 68. *seech.* Seek.

l. 74. *swink due to their hire.* Labour which is their responsibility.

27 l. 80. *ounce.* Lynx.

29 l. 133. *foen.* Foes.

l. 148. *But yet, O man.* This has generally been seen to have a political application, though opinion has been divided as to whether Sidney endorses rebellion against tyranny. According to Ringler (413) 'the general moral is clear—a powerful aristocracy is the best safeguard of the common people against tyranny'. Later commentators have not found the message so clear, and have discussed in detail the relation of Sidney's ideas to those of Languet and other Calvinist political thinkers (cf. Martin Bergbusch, 'Rebellion in the *New Arcadia*', *PQ* 53 (1974), 29–41; Richard McCoy, *Rebellion in Arcadia* (1979): Martin N. Raitiere, *Faire Bitts: Sir Philip Sidney and Renaissance Political Theory* (Pittsburgh, 1984). However, the political application of the fable, whatever it may be, has force only if the plea for tenderness towards animals is also accepted on a literal level. Exceptionally for an Elizabethan, Sidney seems to have disliked hunting; according to Sir John Harington he 'was wont to say, that next hunting, he liked hawking worst' (*A New Discourse of a Stale Subject* [1596], ed. E. S. Donno (1962), 108).

l. 149. *gloire.* Glory.

l. 154. *or know your strengths.* William R. Drennan points out that Sidney's friend Fulke Greville used this phrase in a speech in the House of Commons in 1593 in a context that makes it clear that it refers to justified resistance to oppression ('Or know your strengths: Sidney's attitude to rebellion in "Ister Banke"', *N & Q* 231 (1986), 339–40). This may resolve some of the uncertainties about the political application of the fable.

30 *Double sestina: 'Ye goat-herd gods'; followed by dizain and crown: 'I joy in grief'.* Various attempts have been made to explain the significance of the gentlemen-shepherds Strephon and Klaius and their lost mistress, Urania (see for instance K. Duncan-Jones, 'Sidney's Urania', *RES* xvii (1966), 124–32; Alastair Fowler, *Conceitful Thought* (1975), 56–8). Whatever their precise significance in Neoplatonic terms, it is clear that these companionable lovers of an unattainable and absent mistress represent a higher kind of love.

31 l. 42. *serene.* Harmful summer dew. Evening and morning dews were thought to bring fatal diseases with them; cf. *Julius Caesar*, II. i. 261–3.

32 l. 67. *she, with whom compared the Alps are valleys.* A hyperbole perhaps not here intended to be ridiculous, as it is in the Red Queen's remark in Lewis Carroll's *Through the Looking Glass*: 'I've seen hills, compared with which this is a valley'.

dizain . . . crown. 'Dizains', or ten-line stanzas ending in couplets, seem to have been first referred to by Gascoigne in *Certain Notes of Instruction* (1575). This appears to be the first English reference to the 'crown', a sequence of stanzas or sonnets in which the first line of each repeats the final line of its predecessor, the last line of all repeating the first. Sidney's brother was to attempt a 'Crown of Sonnets' (Robert Sidney, 174–81; cf. also Donne's 'La Corona').

34 l. 63. *the fish torpedo fair.* In *NA* (367) the torpedo fish, or electric ray, is used as an *impresa* on his shield by the accident-prone Amphialus.

l. 72. *crowned basilisk.* A mythical crowned serpent whose gaze was fatal; cf. Pliny, *Natural History*, viii. 33.

l. 80. *spent.* Destroyed.

35 l. 103. *stroys.* Destroys.

Pastoral elegy: 'Since that to death'

36 l. 29. *ai.* The letters imagined by Greek poets as inscribed on the hyacinth after the metamorphosis of the young Hyacinthus; cf. Moschus, *Elegy on Bion*, v. 5 ff.; Ovid, *Metamorphoses*, X. 215; and Milton's 'Lycidas', 106, 'that sanguine flower inscrib'd with woe'.

l. 61. *Philomela.* The nightingale, imagined as pricking her breast against a thorn; cf. *CS* 4.

37 l. 86. *produce.* Draw out, extend.

38 l. 106. *O elements, by whose (they say) contention.* Renaissance commonplace; cf. Marlowe, *Tamburlaine*, I, II. vii. 18–20.

l. 111. *Atropos.* The Fury who cuts off the thread of man's life.

l. 116. *Aesculapius.* The god of medicine.

l. 127. *in turn of hand.* 'In the twinkling of an eye'.

39 *Farewell O sun.* The first stanza was quoted by C. S. Lewis as an example of 'Golden' poetry, but his use of the 1593 text, in which 'woeful's' in line 4 appears as 'joyful's', led him to dismiss the line as 'vapid' and the whole poem as 'empty': a striking example of the need for sound texts as a foundation for criticism (*English Literature in the Sixteenth Century* (1954), 326–7).

l. 8. *queint.* Quenched.

41 *Astrophil and Stella.* Written, probably, between 1 November 1581 (when Penelope Devereux, on whom 'Stella' is modelled, married Robert, Lord Rich) and the end of 1582 (Ringler, 438–9). However, the sequence may incorporate poems or sonnets written earlier, and Ringler's argument that 'it is scarcely probable' that the composition of the sequence extended into 1583 or later because 'it was carefully planned' is not conclusive. If we knew the nature of 'this great cause'

(*AS* 107. 8) in the penultimate sonnet we might be in a better position to suggest a terminal date. It may be reasonable, however, to conjecture that *AS* was completed before 1 Sept. 1583, when Sidney married Frances Walsingham, daughter of the Secretary of State. There is no evidence that the title is authorial. It derives from the first printed text, the unauthorized quarto edition published by Thomas Newman (1591). Newman may also have been responsible for the consistent practice in early printings of calling the lover persona 'Astrophel'. Ringler emended to 'Astrophil' on the grounds of etymological correctness, since the name is presumably based on Greek *aster philein*, and means 'lover of a star'; the 'phil' element alluding also, no doubt, to Sidney's Christian name. Some, but by no means all, of the writers close in time and association to Sidney also spell it thus: e.g. Matthew Roydon, Thomas Watson, Gabriel Harvey, and Sir John Harington. *AS* is the first sonnet sequence in English, and formed the model for the many that were to follow during the 1590s. However, few of its successors approached its range, coherence, and variety.

41 **1** Sidney opens the sequence with a metrical innovation (in English): a sonnet in alexandrines, or twelve-syllable lines. Lines 1–4 are quoted by Fraunce (39) as an example of 'Climax . . . a reduplication continued by divers degrees and steps, as it were, of the same word or sound'. l. 2. *she* (*dear she*). Metre, and Sidney's habitual use of parenthesis, support this reading from the 1598 folio, rather than 'the dear she', the reading of the quartos and MSS, adopted by Ringler.

2 l. 1. *dribbed.* Ineffectual, random. l. 3. *known worth.* Whereas Dante, Petrarch, and most of the French Petrarchizers described love at first sight, Sidney denies it, exploiting this to the greater glory of Stella. It is not clear whether Sidney met Penelope Devereux before her arrival at Court in the autumn of 1581, but he might have known of her 'worth' before meeting her, and her father's dying wish in 1576 for a match between his daughter and Sir Henry Sidney's son makes it likely that he did. Penelope was then 13 (Wallace, 169, 244–5; cf. also *AS* 33). l. 10. *slave-born Muscovite.* The Russians—Slavs—were believed by the Elizabethans to enjoy the oppressive rule of their tsar, at this time Ivan the Terrible. They were also thought comically clumsy and barbaric, as in the Muscovite disguise of the four lovers in *Love's Labour's Lost*, V. ii, which probably alludes to this sonnet. l. 14. *paint.* Probably in *OED* sense 5, 'To give a false colouring or complexion to'.

41–2 **3** ll. 1–8. Sidney catalogues four current ways of ornamenting or elaborating verse: invocation of Muses; imitation of Pindar and other Greek poets, as professed by Ronsard and the other Pléiade writers; logical and rhetorical elaboration, introduced into English notably by

Thomas Watson in his *Hekatompathia* (1582); and the use of exotic similes from natural history, initiated in English prose by Lyly, but already employed in poetry by Petrarch and his European followers. Sidney himself uses all four kinds of elaboration in *OA* poems; rhetorical and logical complexity is the only one used persistently in *AS*.

42 **4** l. 5. *old Cato's breast.* Cato the Censor, who lived to the age of 85, was known as 'a bitter punisher of faults' (238). l. 8. *thy hard bit.* Astrophil is imagining himself to be a horse: cf. *AS* 49 and the opening passage of *DP*.

43 **5** l. 11. *Which elements with mortal mixture breed.* Refers to Plato's theory that mortal beauty, clothed in the physical elements, is only a shadow of absolute virtue; the elements combine in a perishable way (cf. *Republic*, x). l. 13. *up to our country.* Cf. Du Plessis Mornay, tr. Mary Herbert, Countess of Pembroke, *A Discourse of Life and Death*: 'Man is from heaven: heaven is his country and his air' (1592), sig. D3^{v}.

6 Another sonnet in alexandrines. ll. 1–4. Reference to Petrarchan paradox and oxymoron; 'wot not what' refers to his repeated phrase '*non so che*', which became the French '*je ne sais quoi*'. 'Freezing fires' are the most often quoted Petrarchan oxymoron; cf. Leonard Forster, *The Icy Fire* (1969). ll. 5–6. Ronsard used Jove's metamorphoses—into a bull for love of Europa, swan for Leda, and shower of gold for Danäe—as metaphors for his love, e.g. in the sonnet translated by Ralegh as 'Would I were changed into that golden shower' (Ronsard, *Amours*, xx; Ralegh, *Poems*, ed. A. Latham (1951), 81–2). But many other poets of the period used them too. ll. 7–8. An allusion to pastoral poetry, possibly in particular to the work of Spenser, whose *Shepheardes Calender* (1579) was dedicated to Sidney. However, the reference might equally well be to numerous Continental pastoral poets, such as Sannazaro in Italy or Marot in France. ll. 9–11. Perhaps a reference to the *dolce stil nuovo* of the Italian poets of the fourteenth century. l. 12. *I can speak what I feel.* Six emphatic monosyllables underlining Astrophil's blunt truth-telling.

7 Penelope Devereux is known to have had dark eyes and fair hair; but there is also a long tradition of praise of black beauty underlying both *AS* and Shakespeare's 'Dark Lady' sonnets. l. 6. *strength:* strengthen.

8 l. 2. *Turkish hardened heart.* Sidney is thinking of Cupid in contemporary Greece, which was part of the Ottoman Empire, and sees him as a refugee from the proverbially cruel Turks; cf. *AS* 30. Cyprus, birthplace of Aphrodite, was taken by the Turks in 1573.

44 **9** ll. 12–14. *Of touch . . . straw.* Elaborate punning metaphor, perhaps

reflecting the influence of recent Spanish poets, playing on three or four senses of 'touch'. Stella's eyes are of touch-stone (black marble), which without 'touching' 'touch', or move, those who see them. The last line introduces yet another reference, to touch-paper or touchwood, which sets the straw, Astrophil, alight.

44 **10** l. 2. *brabbling.* Quarrelling, quibbling.

45 **11** l. 11. *pit-fold.* 'Pitfall, a trap for birds' (Ringler).

12 l. 2. *day-nets.* 'A net used by day in daring [= fascinating] larks or catching small birds' (*OED*). l. 11. *got up a breach*: broken into the enemy's defences.

45–6 **13** ll. 1–6. The coats of arms refer to amorous exploits: Jove's, to his taking the shape of an eagle to carry off the fair youth Ganymede; Mars's, to his love affair with Venus. l. 3. *eagle sables.* Black eagle. l. 4. *talents.* Talons. l. 6. *vert field.* Green field. l. 11. *Where roses gules are borne in silver field.* Possibly a reference to the Devereux arms, *argent, a fesse, gules in chief three torteaux*, or three red discs on a silver background, as well as to Stella's rosy cheeks. Sidney refers to his own crest in *AS* 65. l. 13. *blaze.* Technical term for spelling out or describing heraldic accoutrements, but also an appropriate word for the activity of Phoebus.

46 **14** First of many sonnets showing Astrophil with an uncomprehending or disapproving friend; cf. 20, 21, 23, 27, 51, 88, 92, and 104. l. 3. *him who first stale down the fire.* Prometheus, who stole fire from heaven and was punished by having his entrails continually torn by a vulture. Cf. *CS* 16a. l. 5. *rhubarb.* Bitter purgative.

15 Another sonnet on contemporary poetic styles; cf. 3, 6. ll. 1–4. *You . . . poesy wring.* Those who rifle classical poets for phrases and images, or imitate other poets who have done the same, gathering stale, not 'sweet', flowers of poetry. ll. 5–6. *You . . . rattling rows.* Poets who use alliteration, mimicked in the phrase 'running in rattling rows', such as those early Tudor poets whom C. S. Lewis called 'Drab'. ll. 7–8. Imitators of Petrarch, of whom there had been many in French, Italian, and Spanish; an English imitator close in time to Sidney was Thomas Watson whose *Hekatompathia* was dedicated to Sidney's enemy the Earl of Oxford. l. 8. *denizened.* Naturalized, i.e. originally foreign.

47 **16** l. 9. *while I thus with this young lion played.* Reference to Greek fable of a shepherd who brought a pet lion cub into his family which when it grew up destroyed his flocks; the story was applied by Aeschylus to Helen of Troy (*Agamemnon*, 717–36).

17 l. 6. *prove.* Test, provoke.

18 l. 1. *With what sharp checks I in myself am shent.* With what sharp rebukes I am inwardly shamed. l. 5. *Unable quit . . . rent.* Unable even to get clear—'quit' is an adverb—of my debt to nature. There may also be a reference to 'quit-rent', the small rent paid by a freeholder in lieu of services. The wider sense is that Astrophil is barely capable of staying alive. ll. 9–10. *my knowledge . . . defend.* Possibly alluding to *OA*, Sidney's 'toyful book', and *DP*, in which, having 'slipped into the title of a poet', he proceeded to defend the art.

48 **19** ll. 9–10. *who fare . . . doth fall.* Commonplace comment on astronomers, deriving from an anecdote about the Greek scientist Thales who fell into a well while gazing at the stars; cf. *DP* 219.

20 l. 6. *level.* Aim. l. 7. *black.* Stella's eye, but also an archery term for black ring surrounding inmost circle on a target.

49 **21** l. 1. *caustics.* Burning substances used in surgery. l. 2. *windlass.* Ambush, ensnare. l. 7. *coltish gyres.* Youthful gyrations, probably with allusion to Plato's image of reason as a charioteer to the passions, identified with horses. l. 9. *mad March great promise made of me.* Probably refers to Sidney's early travels and reputation in Europe, and in particular his embassy to the Holy Roman Emperor in the spring of 1577.

22 l. 2. *from fair twins' golden place.* The zodiacal sign of Gemini, which the sun leaves in late June. l. 5. *by hard promise tied.* Suggests that the ladies rode out to keep an important appointment, and may indicate, *pace* Ringler (468), that the sonnet is based on a real incident.

50 **23** l. 5. *how my spring I did address.* Cf. *AS* 21.9–10; perhaps alludes to Sidney's studiousness during his three years of European travel. l. 7. *the prince my service tries.* The monarch makes use of me. Sidney had been appointed Royal Cup-bearer in 1576 (Wallace, 165), but the allusion here may be to some more significant court duty. l. 9. *harder judges judge ambition's rage.* Still a concern of Sidney's in the last year of his life, when he wrote to his father-in-law: 'I understand I am called very ambitious and proud at home, but certainly if they knew my heart they would not altogether so judge me'.

24 Unlike the miser, ll. 1–8, who at least appreciates the value of his gold, the possessive husband may subject his wife to 'foul abuse'. Presumably an invective against Penelope Devereux's husband, Lord Rich, though J. G. Nichols has suggested that Sidney and Lord Rich may actually have been 'very friendly', and the attacks on him no more than rough banter (*The Poetry of Sir Philip Sidney* (1974), 96).

25 ll. 1–4. *the wisest . . . would raise.* Plato, scholar of Socrates, who had been adjudged wisest of men by the Delphic Oracle ('Phoebus' doom'), did indeed say that if we could see the true form of virtue we should instinctively love it (Plato, *Apology,* 21). But Sidney may be deriving Plato's idea from Cicero (*De officiis,* 1.15), as suggested by the allusion in *DP* (122) to 'the saying of Plato and Tully . . . that who could see virtue would be wonderfully ravished with the love of her beauty'.

51 **26** A light treatment of a weighty Renaissance question: whether the stars were placed simply for our delight, or, as in the medieval view, as pervasive influences on human character and behaviour. Sidney, as distinct from Astrophil, was probably one of the 'dusty wits'. He had a horoscope cast in 1570, perhaps at the behest of his uncle, Leicester, but seems not to have taken the trouble to take it away from Oxford (Bodleian MS Ashmole 356 (5); Osborn, 517–22). Two of his *imprese* celebrate the sheer beauty of the stars: Phalantus's starry shield with a motto 'signifying that it was the beauty which gave it the praise' (*NA* 94), and a pennon carried in his funeral procession showing a fish gazing up at the stars with the motto *Pulchrum propter se* (Lant, *Roll*, 6). Moffet (*Nobilis*, 75) recorded that Sidney particularly disliked judicial astrology, though his anecdote of the 3-year-old Sidney worshipping the new moon (70) confirms his appreciation of celestial beauty.

27 l. 9. *Yet pride, I think, doth not my soul possess.* Sidney, like Astrophil, was conscious of being thought proud; cf. *AS* 23.9 and note. l. 12. *overpass.* Ignore.

28 This address to laborious interpreters of his poems to 'Stella' suggests that Sidney imagined an audience for his sequence.

52 **29** A complex military metaphor is applied to Stella in ll. 1–12: she allows Cupid to keep arms in every part of her body except her heart, so that the heart itself, her capital city, may be kept free of love. This is succeeded in ll. 13–14 by the idea that Astrophil, simply because he has looked at her outward beauty, has been taken captive by love for her.

30 The seven topical questions to which Astrophil is indifferent place the sonnet in the summer of 1582 (Ringler, 470–1). ll. 1–2. The Turks were a threat to Western Europe well into the seventeenth century, and in the early summer of 1582 an attack on Spain was expected. ll. 3–4. Stephen Bathory, the elected king of Poland, invaded Muscovy (Russia) in 1580 and besieged Pskov until Dec. 1581. By the summer of 1582 a treaty had been signed, but Sidney may not have heard of this. l. 5. The three parts are the Catholics, the Huguenots, and the moderate Politiques, who struggled for control of France until the accession of Henri de Navarre in 1589. l. 6. A reference to the Germans (Deutsch), not Dutch (who come in the next line). The Diet of the Holy Roman Empire was held at Augsburg from early July to Sept. 1582 (cf. paper by E. G. Fogel referred to by Ringler, 471). ll. 7–8. The towns of Breda, Tournay,

Oudenarde, Lier, and Ninove were won by the Spaniards during 1581–2; the hope of the Dutch lay in William of Orange. ll. 9–10. Sir Henry Sidney subdued the province of Ulster during his third term of office as Lord Deputy Governor of Ireland, 1576–8, partly by dividing it into shires, and partly by imposing a 'cess', or land-tax, on the great lords; this may be referred to in the 'golden bit'. l. 11. The confusion of the political situation in Scotland is suggested by the word 'weltering' (editions of *AS* during James I's reign tactfully emend to 'no weltering'). During the summer of 1582 there were complex intrigues leading up to the Raid of Ruthven on 22 Aug.

53 **31** l. 14. *Do they call virtue there ungratefulness?* Inversion of the order of subject and object makes it hard to determine whether this means 'Do ladies in heaven call their lovers' virtue "ungratefulness"', i.e. 'unpleasingness', or 'Do ladies in heaven call their own ungratefulness virtue?' The second sense is the likelier, however; and cf. *AS* 5.42 and note.

32 ll. 1–2. *Morpheus . . . living die.* Morpheus, son of Somnus, had the special function of bringing human images to dreamers (Ovid, *Metamorphoses*, xi. 735). Sidney may be particularly recalling Chaucer's *Book of the Duchess*, in which the dreamer reads of Morpheus bringing the drowned King Ceyx to his wife Alcyone, and subsequently himself has a dream of death and living death. l. 3. *an history*: presumably 'a story teller', though not in *OED* in this sense. l. 9. *of all acquaintance.* Probably, 'for friendship's sake'.

33 Somewhat obscure, but presumably alludes to the abortive scheme to betroth Sidney to Penelope Devereux in 1576 (cf. *AS* 2 note). It makes better sense if we believe, with Wallace (156), that Sidney had seen her some years before her marriage to Lord Rich, rather than, with Ringler (436 ff.), that he first saw her about the time of her marriage in 1581: 'rising morn' would then refer to Penelope's appearance as a child of 12 or so, 'fair day' to her fully developed beauty. The meaning of the last line is probably: 'Would that I had been foolish enough to fall in love with Stella when I first saw her, or wise enough never to fall in love at all'.

54 **34** Dialogue between the passionate Astrophil and his 'wit', or reason. l. 4. *oft cruel fights well pictured forth do please.* Refers to Aristotle's *Poetics;* cf. *DP* (227). l. 7. *fond ware*: 'foolish trifles' (Ringler). l. 8. *close.* 'Kept private, not allowed to circulate' (Ringler).

35 l. 4. *nature doth with infinite agree.* 'Stella, though a product of finite nature, is goddess-like and therefore infinite' (Ringler). l. 5. *Nestor's counsels.* Nestor was the praeternaturally aged counsellor who gave advice to the Greeks in Homer's *Iliad.*

55 **36** l. 2. *yelden.* Yielded. ll. 12–14 *not my soul . . . from thee.* Not only Astrophil's soul, which has sense, but stones and trees, which lack it,

are enchanted by Stella's voice. Astrophil had already been conquered through one sense, that of sight; now he is conquered through another in hearing her singing voice. For evidence that Lady Rich was indeed musical, see 'Lady Rich', 185–8. There is an implicit allusion to Orpheus, whose power to enchant stones and trees was described in Ovid, *Metamorphoses*, x. 11; cf. also *AS* 3.1–6

55 **37** Appears in only one MS, and in neither of the quarto editions; first printed in 1598 folio. It may have been suppressed from circulated texts because the attack on Lord Rich was too explicit. Though the language is distanced and romance-like—as in 'lordings' —the allusions are clear. The seat of the Riches was in Eastern England, Leighs, in Essex, which is presumably indicated by the fairy-tale-like phrase 'Towards Aurora's court'; there may also be a sense of 'near Queen Elizabeth's court', though Aurora was not one of her usual names. The 'riddle' posed by the nymph's 'misfortune' would pose little difficulty to contemporary readers aware of Penelope Devereux's forced marriage, at which, it was later said, 'she did protest at the very solemnity and ever after' (Wallace, 247).

38 l. 2. *hatch.* Close. *unbitted.* Unrestrained. l. 5. *error.* Wandering. l. 7. *so curious draught.* Such painstaking workmanship, or drawing. l. 8. *not only shines, but sings.* Cf. Strephon's vision of Urania in *Lamon's Tale.* l. 14. Stella has murdered sleep; cf. *Macbeth*, II. ii. 35.

56 **39** l. 5. *press.* Crowd. ll. 9–14. *Take thou . . . weary head.* The offer of gifts to Morpheus is conventional, but a specific source may be Chaucer's *Book of the Duchess*, 240–69; cf. *AS* 32 note. The 'rosy garland' probably means a garland of secrecy, as in the phrase *sub rosa.*

57 **40** l. 14. *thy temple.* Sidney's brother was also fond of the image of the lover's heart as a temple to his lady's love: cf. Robert Sidney, 139, 183, 207.

41 Probably refers to *Triumph* (299–311). Though Sidney took part in other tournaments in 1581–2, this was much the most striking one at which a French delegation was present. l. 6. *daintier.* 'More precise' (Ringler). l. 7. *sleight, which from good use doth rise.* Dexterity achieved by plenty of practice. l. 11. *nature me a man of arms did make.* Sidney's father and grandfather, Sir Henry and Sir William Sidney, were both tilters in their youth; so were his mother's brothers, the Earls of Leicester and Warwick (Young, 126–7, and *passim*). The phrase 'man of arms' refers specifically to one who fought in the semi-medieval armour of the tilt; cf. the description in *NA* (91) of Phalantus as 'the fair Man-of-Arms'.

42 l. 14. *Wracks triumphs be, which love (high set) doth breed.* Cf. Petrarch. *Canzoniere*, 140. 14, translated by Surrey as 'Sweet is his death, that takes his end by love'.

58 **44** l. 7. *overthwart*. 'Perverse, contrary' (Ringler).

45 l. 3. *cannot skill*. Is not able. l. 14. *pity the tale of me*. A use of the Aristotelian paradox that objects represented in art may have an emotive effect that they lack in life. *AS* as a whole may constitute 'the tale of me'.

59 **47** l. 2. *burning marks*. 'Brands indicating slavery' (Ringler).

60 **48** l. 14. *A kind of grace it is to slay with speed*. The French *coup de grâce* may underlie this; cf. also Petrarch, '*Un modo di pietate, occider tosto*' (Ringler).

49 There are parallels to the image of the lover as a horse ridden by love in Petrarch (Ringler, 476); it has a particular appropriateness to a poet whose first name means 'horse-lover', and who was elsewhere almost persuaded 'to wish myself a horse' (*DP* 212). l. 7. *boss*. Metal knob on the bit. l. 14. *manage*. Technical term for the movements of a highly trained horse. Astrophil presumably takes delight both in the horse's 'manage' and in his own, guided by Cupid.

50 A self-sustaining artefact: Stella's name, which opens the sonnet, preserves it from deletion. Perhaps imitated by Shakespeare in *The Two Gentlemen of Verona*, I. ii. 195–30, where Julia shows tenderness to the name 'Proteus' written on scraps of a letter she has just torn up. l. 8. *portrait*. Portray. l. 11. *babes*. His newborn thoughts.

61 **51** Apparently addressed to a fellow-courtier whose solemn talk of politics, court intrigues, and quests for favour Astrophil finds tedious and incongruous with his pleasant reflections on Stella. l. 5. *silly*. Innocent, naïve. l. 7. *in steed*. Instead. l. 10. *most troubled streams*. An image often used for the flowing channels of the Queen's bounty; cf. Ralegh:

> Those streams seem standing puddles, which before
> We saw our beauties in, so were they clear.
> Belphoebe's course is now observed no more.
>
> (*Poems*, ed. A. Latham (1951), 34).

52 'Love asserts his right of possession ("title") to Stella by citing facts—that she outwardly wears his badge or livery; Virtue enters a demurrer—admits the facts but denies that they establish legal title by raising the question whether the essential Stella is her inside or her out; this stops the action (stays the suit) until the court can decide the legal point' (Ringler).

61–2 **53** Contrast with *AS* 41; whether Astrophil does well or badly in the tiltyard, he attributes the outcome to Stella. l. 2. *staves*. Plural of (tilting) staff. The object of tilting was to break one's staff on the opponent's shield. l. 11. *to rule*. To control (the horse). l. 12. *trumpet's sound*. A trumpet signalled the beginning of each fresh course in the tilt.

62 **54** l. 13. *Dumb swans, not chattering pies, do lovers prove.* Quoted by Nashe in *Summer's Last Will and Testament* (1592/3):

> Well sung a shepherd that now sleeps in skies
> 'Dumb swans do love, and not vain chattering pies'.
> (*Works*, ed. McKerrow (1958), iii. 271).

55 l. 8 *How their black banner might be best displayed.* A military version of the paradox expressed in *AS* 2.14, 'While with a feeling skill I paint my hell'. Astrophil's sad words are led by a black banner both because they are written in black ink and because they are associated with destruction (cf. Marlowe, 1 *Tamburlaine*, IV. i. and *passim*).

63 **56** l. 3. *a whole week without one piece of look.* If it is remarkable that Astrophil has had a week without sight of Stella we may perhaps imagine a good many habitual encounters between the two. l. 11. *phlegmatique.* Cold and moist; spelling retained for the sake of metre.

57 ll. 10–11. *But them . . . darkness clear.* Probably the suggestion is that Stella sings the actual sonnets of the sequence we are reading, and in so doing sweetens them and drains them of pathos.

63–4 **58** l. 1–8. *Doubt . . . rudest brain.* Refers to a controversy in classical oratory about whether the words or the manner of delivery have more influence on the audience (Cicero, *De oratore*, ii. 223; Quintilian, XI. xi. 2–4). The image of rhetoric's golden chain is traditional, employed, for instance, in the emblem in which Hercules is shown leading crowds by golden chains from his mouth (Alciati, *Emblemata* (Antwerp, 1574) 458).

64 **60** l. 1. *my good angel.* Refers to the classical theory that every human being is accompanied by two spirits, one good, one bad, which prompt him accordingly; cf. Shakespeare, *Sonnets* 144, and the Good and Bad Angels in Marlowe's *Doctor Faustus.*

65 **61** l. 7. *selfness.* Selfishness, egotism (a coinage of Sidney's).

62 l. 5. *thus watered was my wine.* so my hopes were moderated; cf. Isa. 1: 22. l. 6. *a love not blind.* Refers to the Neoplatonic theory of two Venuses and two Cupids, earthly and heavenly (Ficino). The earthly Cupid, or Desire, is blind; the heavenly, or Platonic, is sighted. The divine Cupid is often shown in emblems (e.g. Alciati, *Emblemata* (Antwerp, 1574), 297).

63 l. 3. *dove.* 'An appellation of tender affection' (*OED*). Astrophil is becoming increasingly possessive in his ways of speaking to Stella. l. 14. *in one speech two negatives affirm.* A rule applying to Latin, not English, grammar, but Astrophil adapts it to his purpose, the question 'to grammar who says nay?' neatly repeating the idea of a negative reply. The refrain of *AS* 4 ('No, no, no, no, my dear, let be') shows that in English doubled negatives may emphasize, not affirm.

67 **First song** l. 22. *Who long-dead beauty with increase reneweth?* Presumably implies that Stella calls ancient beauties, such as Helen or Laura, back to life, but is even more beautiful than they. l. 32. *not miracles are wonders.* Wonders are not miracles; i.e. wonders in you appear as a matter of course, they are not contrary to nature.

64 l. 1. *no more these counsels try.* Stella seems to have been talking to Astrophil, trying to dissuade him from love.

68 **65** ll. 5–8. *For when . . . mine eyes.* The fable of the runaway Cupid given refuge ('harbour') by the lover derives from the *Anacreontea*, 33, imitated also by Greville, *Caelica*, 12. l. 14. *Thou bear'st the arrow, I the arrow head.* Refers to the Sidney arms, *or, a pheon azure*, a blue arrow head on a gold background.

66 l. 8. *stilts.* Crutches.

67 l. 8. *What blushing notes dost thou in margin see?* The 'text' is Stella's eyes, the 'margin' her cheeks, blushing, as in the preceding sonnet.

69 **68** *Amphion's lyre.* Thebes was built with his music; cf. *AS* 3.4. l. 14. *to enjoy.* With sexual undertone.

69 ll. 7–8. *Gone is the winter of my misery | My spring appears.* Perhaps echoed by Shakespeare, *Richard III*, 1. i. 1–2.

70 **70** l. 4. *I Jove's cup do keep.* Ganymede mixed nectar for Jove, and Astrophil is promoted to heavenly favour by Stella's promise; it may also be relevant that Sidney held the office of Royal Cup-bearer.

71 l. 7. *night-birds.* 'Vices (the owl, for instance, is used variously to pictorialize avarice, envy, sloth, and gluttony)' (Ringler). l. 14. T. P. Roche has pointed out that this line, a vivid expression of the reason why Astrophil's virtuous covenant with Stella will not hold, comes at the numerical midpoint of the sequence, if all the lines are counted ('*Astrophil and Stella*: A Radical Reading', *Spenser Studies*, 3 (1982), 144, 186).

72 l. 8. *Virtue's gold now must head my Cupid's dart.* Cf. *AS* 65.14; the blue arrow head of the Sidneys must now be tipped with gold.

71 **Second song** Makes Astrophil's desire more explicit than in the preceding sonnets, where it is to some extent veiled in double meaning (cf. *AS* 68.14). Stella's awakening and anger prevent seduction; cf. *OA* (202), where Musidorus's assault on the sleeping Pamela is interrupted by 'a dozen clownish villains'. l. 14. *Cowards love with loss rewardeth.* Love rewards cowards with loss, i.e. 'None but the brave deserve the fair'. l. 26. *Louring.* Scowling.

72 **73** l. 4. *so soft a rod.* The worst Cupid has to fear is an angry look from his mother Venus. l. 11. *Those scarlet judges, threatening bloody pain.* Stella's lips, like High Court judges, are robed in scarlet; Astrophil's offence is serious.

72 **74** l. 1. *Aganippe well.* Fountain in Greece dedicated to the Muses. l. 2. *Tempe.* Valley in Thessaly where Apollo pursued Daphne and she was changed into laurel. ll. 5–6. *Some . . . mean by it.* Consistent with Sidney's dismissal of the idea of poetic inspiration in *DP* (240). l. 7. *blackest brook of hell.* Styx; cf. *OA* 73.161 and *CS* 17.6. l. 11. *my verse best wits doth please.* Astrophil does not seem to have reserved his sonnets for Stella's eyes only.

72–3 **75** Apparently oddly placed among the group of sonnets reflecting on Astrophil's stolen kiss and its aftermath, this sonnet survives in a MS belonging to Sir John Harington (H. R. Woudhuysen, '*Astrophel and Stella* 75: A "new" text', *RES* 37 (1986), 388–92). l. 4. *imp.* Graft. ll. 5–8. *Nor . . . did late obtain.* Edward IV usurped the throne in 1461, after his father, the Duke of York, was killed while fighting against the Lancastrians: the later years of his reign were tranquil and well ordered. ll. 9–11. *Nor that . . . a tribute paid.* In 1474 Edward invaded France, and was persuaded by Louis XI to withdraw with a payment of 75,000 crowns. France was not, in fact, 'hedged', or protected, at this time by the 'bloody lion' (Scotland, represented a red lion). l. 14. *To lose his crown, rather than fail his love.* In 1464 Edward married Lady Elizabeth Grey, widow of Sir Richard Grey. Warwick, the 'King-maker', who had been negotiating a French match for him, drove Edward into exile in 1470, though he recovered his throne in the following year. Cf. Shakespeare's *3 Henry VI*, III. ii and *passim.*

73 **76** Another sonnet in alexandrines. l. 13. *walking.* With agitated thoughts, mind 'racing'.

77 In alexandrines. l. 2. *whose lecture.* The reading of which.

74 **78** A characterization of the monster Jealousy, comparable with the monster Cupid anatomized in *OA* 8. It is presumably directed towards Lord Rich, as suggested in the final wish that he should be cuckolded.

79 ll. 1–2. *Sweet . . . sweetest sweetener art.* A use of a 'polyptoton', when a single word with 'divers fallings or terminations' is frequently repeated (Fraunce, 51–2).

75 **80** l. 8. *honour's grain*: 'The purple of political dignity' (Ringler). l. 12. *resty.* Restive.

82 l. 1. *Nymph of the garden.* The garden is Stella's own body; the cherry tree is her lips. Cf. Thomas Campion's song 'There is a garden in her face' (*Works*, ed. W. R. Davis (1967), 174). l. 3. *His who till death looked in a watery glass.* Narcissus drowned while gazing at his own reflection in a river. l. 4. *hers whom naked the Trojan boy did see.* Paris saw Venus naked.

76 **83** l. 1. *Good brother Philip.* Stella's sparrow is 'brother' to Astrophil both because they are rivals for her favour and because they share a

Christian name. The sparrow, traditionally identified with lechery, is partly an image for Astrophil's own desire; cf. Gascoigne's *The praise of Phillip Sparrowe*, where the innuendo is inescapable (*Posies*, ed. J. W. Cunliffe (1907), 455–6). Other sources are Catullus's elegy on Lesbia's sparrow and Skelton's *Philip Sparrow*. l. 3. *your cut to keep*. To behave with modesty, restraint. l. 14. *sir Phip*. 'Sir' is used here 'With contemptuous, ironic or irate force' (*OED* 6b). Sidney's knighthood, which he received in Jan. 1583, is unlikely to be relevant.

76 **Third song** l. 1–4. For Orpheus and Amphion, cf. *DP* (217). ll. 7–10. *If love . . . endless night*. Alludes to two anecdotes in Pliny's *Natural History* (viii. 61, x. 18). Thoas, an Arcadian, was rescued from robbers by a dragon (lizard) to which he had been kind; and a maiden of Sestos nurtured an eagle which loved her so much that it flew into her funeral pyre and was consumed.

77 **84** Addressed to the road leading to Stella's house. Mona Wilson suggested that this was the Whitechapel Road, leading to Wanstead (*Sir Philip Sidney* (1950), 191); but the Rich seat, Leighs, in Essex, or Penelope Devereux's brother's house, Essex House, in London, seem slightly more likely destinations. l. 2. *to some ears not unsweet*. Another hint that these poems are being read and appreciated by readers other than Stella; cf. *AS* 74.11. ll. 3–4. *Tempers . . . chamber melody*. According to a great-uncle of John Aubrey's, Sidney 'was often wont, as he was hunting on our pleasant plains, to take his table book out of his pocket and write down his notions as they came into his head' (Aubrey, *Brief Lives*, ed. A. Clark (1898), ii. 248).

85 ll. 3–4. *joy . . . strain*. In *OA* (229) Pyrocles, on his way to Philoclea's chamber, 'found that extremity of joy is not without a certain joyful pain, by extending the heart beyond his wonted limits'. l. 6. *Not pointing to fit folks each undercharge*. Not assigning secondary duties to appropriate underlings. l. 14. *Thou but of all the kingly tribute take*. Astrophil's heart, the addressee here, should take possession of Stella's.

78 **Fourth song** l. 15. *These sweet flowers on fine bed too*. Presumably flowers embroidered on a bedspread, since the lovers are indoors. l. 37. *Your fair mother is abed*. Penelope Devereux's mother, Lettice Knollys, had been married to the First Earl of Essex (d. 1576); she then married Sidney's uncle, the Earl of Leicester, in 1578, thus making Sidney a step-cousin of her children. She seems to have been at least as remarkable a personality as her daughter, whom she outlived by nearly thirty years. l. 39. *you do letters write*. Unlike many less well educated Elizabethan ladies, Penelope Devereux was a fluent and accomplished letter writer ('Lady Rich', 188–9).

79 **86** l. 5. *like spotless ermine*. The ermine was believed to die rather than allow its skin to be spotted; cf. *OA* 62.116 (136).

79–82 **Fifth song** Astrophil punishes Stella for her rejection of him by offering her a clumsy and old-fashioned poem ('Lady Rich', 178). l. 31. *Your client poor my self, shall Stella handle so?* Can Stella, exalted by Astrophil's muse, get away with treating him so badly? l. 36. *Sweet babes must babies have.* Nice children must have dolls. l. 42. *Ungrateful who is called, the worst of evils is spoken.* 'Ingratitude comprehends (is the worst of all) faults (vices)' (Tilley, 166). l. 46. *rich in all joys.* No doubt another pun on Stella's married name. ll. 89–90. *such skill in my muse . . . shall be proved.* If Stella will only be kind to Astrophil, he will again write skilful poetry in her praise, instead of the present inelegant abuse.

82 **Sixth song** Included in William Byrd's *Psalms, Sonnets and songs of sadness and piety* (1588). l. 4. *former.* First, higher. l. 43. *common sense.* Shared or community feeling or judgement (*OED*, sense 3).

84 **Seventh song** Quoted by Fraunce (46) as an example of 'Epanodos, regression, turning to the same sound, when one and the same sound is repeated in the beginning and middle, or middle and end'.

Eighth song Imitated by many later poets; e.g. Fulke Greville, *Caelica* 75; Edward, Lord Herbert of Cherbury, 'Ode upon a question moved: whether love should endure for ever'. There is a setting by Robert Dowland in *A Musical Banquet* (1610). l. 1. *In a grove most rich of shade.* The suggestion may be that the shade is made rich by Lady Rich's presence, and not necessarily that the meeting occurs in the garden of one of Lord Rich's houses. l. 19. *arms crossed.* The pose of melancholy, or like effigies on tombs. l. 104. *my song is broken.* The attempt at objectivity represented by third person narration collapses here; cf. the complementary shift from first to third person in the following song, ll. 26–30.

87 **Ninth song** There is a setting by Robert Dowland in *A Musical Banquet* (1610): also in Christ Church, Oxford, MS 439, fo. 9. l. 23. *caitiff.* Wretched, miserable.

89 **87** l. 4. *iron laws of duty.* Not identifiable. l. 5. *she with me did smart.* She shared my pain.

88 Astrophil is tempted to unfaithfulness by the flirtatious approaches of another lady, but maintains his devotion to Stella through inward recollection. l. 6. *sun.* Consistent with identification of Stella elsewhere with the sun rather than, as her name might suggest, a star. l. 8. *cates.* Choice victuals.

90 **89** Only two 'rhyme' words are used, 'night' and 'day', but in a pattern approximating to the Petrarchan form. Fraunce (48) quoted ll. 5–14 as an example of '*epanodos*, regression'.

90 Perhaps Stella has accused Astrophil of using her as a pretext for enhancing his own reputation through poetry; this would be consistent with other hints, in this part of the sequence, of incipient disintegration of the relationship. l. 10. *laud*. Praise, credit.

91 **91** l. 1. *honour's cruel might*. Presumably the same as the 'iron laws of duty' which called Astrophil away in *AS* 87.4. l. 11. *Models such be wood-globes of glistering skies*. 'Wood-globes' must be globes showing the constellations, which were in use throughout the sixteenth century. Sidney studied the 'sphere' during his year in Italy, and he recommended its study to his brother Robert and to Edward Denny in 1580 (290).

92 l. 1. *Indian ware*. Ware from the Indies, proverbially scarce and expensive. In *The Defence of Leicester* Sidney says that, given his noble ancestry, Leicester's want of gentility 'would seem as great news as if they came from the Indies' (*Misc. Prose*, 133). l. 3. *cutted Spartans*. The Spartans were notorious for their 'cutted', or concise, pithy, mode of rhetoric. l. 5. *total*. Brief.

Tenth song Set by William Byrd in *Songs of sundrie natures* (1589) and by Robert Dowland in *A Musicall Banquet* (1610). ll. 19–48. *Thought . . . nectar drinking*. Fulke Greville's *Caelica*, 45. 41–50, seems to comment on this passage in which Astrophil sends his 'Thought' off on a journey of sexual exploration:

But thoughts, be not so brave
With absent joy;
For you with that you have
Yourself destroy.
The absence which you glory
Is that which makes you sorry
And burn in vain:
For thought is not the weapon
Wherewith *thought's-ease* men cheapen:
Absence is pain.

('Lady Rich' 191)

93 **93** ll. 6–8. *my foul stumbling . . . did miss*. The exact nature of Astrophil's offence is as obscure as the 'fault' of the young man in Shakespeare's *Sonnets*; but he seems to claim that he has been trying too hard to do or say the right thing. l. 11. *'quit*. Acquit.

94 **96** l. 9. *mazeful*. Bewildering.

97 The full moon (Diana) accompanied by stars makes a brilliant night, but the night would prefer the sun (Phoebus); likewise, another, unidentified lady ('Dian's peer', or equal) tries to console Astrophil for the absence of Stella (the sun). l. 4. *standing*. Shooting position (archery term). l. 8. *dight*. Dressed.

95 **98** As Ringler says, Addresses to the bed were frequent in Renaissance

love poetry', but 'Sidney's handling is, as usual, entirely original'. l. 4. *lee shores.* Shores facing away from the wind. l. 12. *when Aurora leads out Phoebus' dance.* When dawn leads in day. l. 13. *wink.* Close.

95 **99** l. 3. *mark-wanting shafts.* Arrows lacking a target. l. 9. *charm.* Sing in unison.

96 **100** l. 1. *O tears, no tears, but rain from beauty's skies.* An example of epanorthosis, or 'correction . . . when anything passed is called back' (Fraunce, 78–9). Perhaps imitated by Kyd, *Spanish Tragedy*, III. ii in the speech beginning 'O eyes, no eyes, but fountains fraught with tears', which in turn was mocked by Jonson in *Every Man in his Humour*, I. v. 57–8. Sidney's own model was probably Petrarch or one of the Pléiade poets who used the device. l. 9. *conserved in such a sugared phrase.* Refers to the preservation of fruits in sugar.

101 l. 11. *prest.* prompt, ready. l. 14. *heaven stuff.* 'Heavenly material, her bodily beauty' (Ringler).

102 Another sonnet in alexandrines; cf. 1, 6, 8, 76, 77. l. 5. *vade.* Disappear. l. 9. *Galen's adoptive sons.* 'Followers of Galen, old-fashioned physicians unaware of the new Paduan school' (Ringler). l. 10. *hackney on.* Ride stumblingly, as on a worn-out horse.

97 **103** l. 2. *with many a smiling line.* Cf. *NA* (7), where blood has 'filled the wrinkles of the sea's visage'. l. 9. *Aeol's youths.* Breezes, sons of Aeolus, god of winds.

104 l. 10. *If I but stars upon my armour bear.* Sidney may have worn starry armour in tournaments. A star-spangled banner is shown in Lant, *Roll* (5), and Nathaniel Baxter describes a posthumous vision of him in blue armour covered with silver stars (*Sir Philip Sidneys Ourània* (1606)). Cf. also *NA* (412) for a description of a knight whose clothes were 'all cut in stars, which made of cloth of silver and silver spangles, each way seemed to cast many aspects'.

98–9 **Eleventh song** First printed in the 1598 folio of Sidney's works. There is a setting by Thomas Morley, *The First Booke of Aires* (1600). l. 27. *such minds.* Such states of mind. l. 42. *Argus' eyes.* Argus was a many-eyed monster set by Juno to watch over Jove's mistress, Io.

99 **105** Astrophil has missed an opportunity of seeing Stella by night, but the circumstances are obscure. 'Dead glass' in l. 3 has been variously identified as Astrophil's eye, a telescope, or his tears; Ringler (490) thinks it is Astrophil's eyes, which were not to blame for not seeing Stella. However, 'dazzling race' may suggest a lantern or torch, as in l. 11; and the Bright MS version of l. 1, 'Unhappy light', suggests that this object may be the main addressee throughout.

106 l. 3. *Bare me in hand.* 'Deceived me, promised me falsely'

(Ringler). l. 5. *dainty cheer*. Comfort, choice provisions. l. 10. *conversation sweet*. A phrase used of Stella in *AS* 77.

100 **107** l. 8. *this great cause*. May refer either to public duty or a literary project, such as the translation of Du Bartas, which Stella herself wishes to see accomplished ('what thy own will attends'). Ringler thinks the reference is to 'public service in general', but the phrase sounds allusive and specific.

108 A revision to Astrophil's previous self-absorbed obsession; perhaps placed as a final snapshot of 'Astrophil' before Sidney directs his 'wit' to the 'cause' mentioned in the previous sonnet. The paradoxes and oxymora make it clear that he can go no further as 'Astrophil'. l. 10. *Phoebus' gold*. Sunlight.

101 *The Defence of Poesy*. Written some time after Dec. 1579, when Spenser's *Shepheardes Calender* was dedicated to Sidney (cf. *DP*, 1337–8). Some stimulus may also have been given by the dedication earlier in the same year of Stephen Gosson's *The schoole of abuse*, a Puritan attack on plays and players for which Spenser claimed that Gosson had been 'scorned' by Sidney (*Two Letters*, 1580). However, Gosson dedicated another book to Sidney a few months later, and was apparently promoted by Sidney's future father-in-law, Sir Francis Walsingham, after another attack on the stage, *Playes Confuted* (1582) (A. F. Kinney, *Markets of Bawdrie: The Dramatic Criticism of Stephen Gosson* (Salzburg, 1974), 43–51). Van Dorsten suggested that *DP* was written while Sidney was completing *OA* during 1580, as 'a summary of his views before the early experiments were over' (*Misc. Prose*, 63), but it may equally well belong to the interval betwen the two *Arcadias*, overlapping, perhaps, with *AS*, as hinted by verbal and metaphoric links and a possible allusion in *AS* 18.10. Shepherd (4) assigned it to 1581–3; the date limits may be even wider, from early 1580 to 1585, when William Temple became Sidney's secretary and prepared his *Analysis* of the work.

Among Sidney's models were Aristotle's *Rhetoric* (which he is said to have translated); Horace's *De arte poetica*; J. C. Scaliger's *Poetices libri septem* (1561); Serranus's edition of Plato (1578); Amyot's preface to his translation of Plutarch's *Lives* (translated by North in 1579); and various Italian theorists such as Minturno (*De poeta*, 1559). As Shepherd (12–16) has shown, the treatise is cast in the form of a classical oration, though designed for silent reading, not delivery. Its title is unstable. As van Dorsten says, 'Sidney may never have contemplated giving his discourse a proper title' (*Misc. Prose*, 69); however, his extraction from *DP* of the phrase 'a . . . defence of . . . poetry' seems editorially dubious, since his copy-text was *The Defence of Poesy*, published by Ponsonby in 1595. Olney published the work under the title *An Apology for Poetry* in the same year, and in his manuscript *Analysis* Temple referred to it as '*tractatio de Poesi*'. In the present edition the Ponsonby title, adopted in the 1598 and later editions of Sidney's works, is restored, along with about twenty readings from his text rejected by van Dorsten.

101 ll. 1–2. *When . . . together*. Edward Wotton (1548–1626) was secretary to the English embassy in Vienna at the court of the Holy Roman Emperor, Maximilian II, while Sidney was there in the autumn of 1574, and travelled back to England with him in the spring of 1575 (Osborn, 235, 307–8, and *passim*). He was one of the pall-bearers at Sidney's funeral, and was bequeathed 'one fee-buck to be taken yearly out of my park at Penshurst' in Sidney's will (*Misc. Prose*, 149, 218).

ll. 2–3. *John Pietro Pugliano . . . stable*. The office of 'Esquire of the stable' was a dignified one.

l. 10. *faculty*. profession.

l. 16. *pedanteria*. Italian word for 'pedantry', or heavy-footed book learning, perhaps quoted from Pugliano's native tongue.

l. 20. *a piece of a logician*. One of his Oxford contemporaries, Richard Carew, testified to Sidney's early skill in disputation (*The Survey of Cornwall* (1602), 102^{v}), and his protégé Abraham Fraunce shared with him an interest in the latest development in Ramist logic (cf. Bodleian MS Rawl. D.345). However, another logician in his entourage, his secretary William Temple, made a logical *Analysis* of *DP* which showed its arguments to be in several respects deficient (see Abbreviations).

ll. 27–8. *having slipped into the title of a poet*. This may refer to *OA*, or to Sidney's lyric poetry in general including the poems in *OA* and *CS*; it can refer to the work which reads most like a 'spontaneous overflow of powerful feelings', *AS*, only if, *pace* van Dorsten, we date *DP* after 1582 (see headnote, above).

l. 34. *is fallen to be the laughing stock of children*. The mid-Tudor period—called by C. S. Lewis 'the Drab Age'—was not a very distinguished phase of English poetry, but even so Sidney is probably exaggerating for effect.

102 ll. 45–6. *the hedgehog . . . host*. Aesop, *Fables*, 184–5.

l. 46. *vipers . . . parents*. Pliny, *Natural History*, X. lxxxii. 2.

ll. 48–9. *Musaeus, Homer, and Hesiod, all three nothing else but poets*. Musaeus, whose *Hero and Leander* was translated by Marlowe (published 1598), was believed by the Elizabethans to be an even earlier poet than Homer, though he actually wrote in the fifth century AD. Hesiod, author of *Theogonia* and *Opera et Dies*, really was ancient, writing in the eighth century BC.

l. 51. *Orpheus, Linus*. The archetypal Greek poet Orpheus was believed in the Renaissance to be the author of the 'Orphic hymns'; Linus was supposedly his teacher, so more ancient still.

102 ll. 57–8. *Amphion . . . Thebes*. Thebes was said to have been built by the power of Amphion's music; cf. *AS* 3.2–3.

ll. 58–9. *Orpheus . . . beastly people*. Sidney may imply that it was more remarkable that Orpheus moved the hearts of barbarous people than that he moved beasts.

l. 60. *Livius Andronicus and Ennius*. Livius Andronicus (d. 204 BC), a Greek, was thought to be the earliest Latin poet; Ennius (239–169 BC) wrote *Annales*, a history of Rome, in verse.

l. 62. *Dante, Boccaccio, and Petrarch . . . Gower and Chaucer*. Sidney's emphasis is on the encyclopaedic character of these five poets, as well as their early date.

ll. 69–71. *Thales, Empedocles, and Parmenides . . . Pythagoras and Phocylides . . . Tyrtaeus . . . Solon*. Seven early Greek thinkers: Thales (fl. 585 BC) was believed to have composed poems on astronomy and physics; Empedocles (fl. 450 BC) wrote poems *On Nature* and *Purifications*; Parmenides (fl. 475 BC) founded the Eleatic school of philosophy; Pythagoras (fl. 530 BC) was believed to be the author of *Aurea Carmina*; Phocylides (fl. 560 BC) wrote gnomic verses; Tyrtaeus (fl. 670 BC) was a lame schoolmaster who inspired the Spartans to victory by his verses; and Solon (fl. 600 BC), the great Athenian legislator, was the supposed author of a lost epic on Atlantis (Plato, *Timaeus*, 20e). Sidney probably knew these fragments of early Greek poetry from Henri Estienne's collection *Poesis Philosophica* (1573), whose preface included eulogies of poetry which may have influenced *DP*.

103 ll. 83–4. *the well ordering of a banquet, the delicacy of a walk, with interlacing mere tales, as Gyges' ring*. Supper arrangements are described in Plato's *Symposium*; an outdoor walk in the *Phaedrus*; and the fable of Gyges' ring of invisibility in *Republic*, ii. 359.

l. 87. *historiographers*. Writers of history.

l. 90. *Herodotus entitled his History by the name of the nine Muses*. Herodotus was not himself responsible for naming each of his nine books after a Muse, but Sidney may not have known this. Valla's translation of Herodotus into Latin was corrected by Sidney's friend Henri Estienne (1566).

ll. 101–2. *In Turkey . . . poets*. Like most Elizabethans, Sidney took a keen interest in the Turks, whose Ottoman Empire extended to Cyprus (cf. *AS* 8) and deep into Hungary. T. Washington's *Navigations into Turkey* (1585, but probably completed four years earlier), translated from Nicolas de Nicolay, was dedicated jointly to Sidney and his father. This included plates depicting both 'law-giving divines' and poets, or minstrels.

103 ll. 102–4. *In our neighbour country Ireland . . . in a devout reverence.* Sidney here uses the power of the Irish bards to support his case for the archetypal nature of poetry, though their activities were actually seen as dangerous and anarchic by his father, as Lord Deputy Governor; cf. John Derrick, *The Image of Ireland* (1581), fo. 2^{r-v}.

l. 106. *areytos*. 'Rhymes or ballads' celebrating ancestral valour, accompanied by dancing and music, performed by New World Indians on Haiti (Peter Martyr, *Decades of the newe worlde or West India* tr. Richard Eden (1555), III. vii).

ll. 112–18. *In Wales . . . long continuing.* Sidney had a particular knowledge of Wales and Welsh antiquity through his father, who was Lord President of the Marches for Wales from 1559, and supported the completion by David Powell of the learned antiquary Humfrey Lhuyd's *Commentarioli Brittanicae (The historie of Cambria* (1584), dedicated to Sidney).

104 ll. 131–5. *whereof . . . performed it.* Julius Capitolinus, *Vita, Albini*, in *Scriptores Historiae Augustae*, tells this anecdote; line from Virgil, *Aeneid*, II. 314, tr. by T. Phaer (1580) as 'With anger wood [= mad], and fair me thought in arms it was to die.'

ll. 140–1. *the oracles of Delphos and Sybilla's prophecies were wholly delivered in verses.* The Pythian priestess at Delphi declared Apollo's oracles in hexameters; the original Sybilline oracles of early Greece hardly survive, but a large body of 'Sybilline' verses was gathered up from late antiquity onwards.

ll. 149–50. *as all learned Hebricians . . . not yet fully found.* Hebrew scholars from Jerome onwards agreed that the Psalms were in verse, and until the work of Lowth in the eighteenth century it was thought that some pattern of quantity or stress or rhyme might be discovered in them.

l. 153. *prosopopoeias*. Personifications.

ll. 154–5. *his telling of the beast's joyfulness and hills leaping*. Ps. 29.

105 l. 169. *a maker*. Obsolete word for 'poet', current in the Renaissance.

l. 196. *Cyclops, Chimeras, Furies*. The Cyclops were one-eyed monsters; Chimeras were winged, goat-like animals; Furies or Erinyes were savage, malign creatures concerned with retribution.

106 ll. 206–9. *so true a lover as Theagenes . . . Virgin's Aeneas*. Theagenes was the young hero of Heliodorus' *Aethiopica*, a late Greek romance (translated by Underdowne, ?1569) which was one of the sources of *OA*; Pylades was the faithful friend of Orestes in Euripides' *Oresteia*; Orlando was the hero of Ariosto's *Orlando Furioso* (1532); Cyrus the young prince whose education was described in Xenophon's *Cyropaedia*;

and Aeneas, hero of Virgil's *Aeneid*, was seen by Renaissance readers as virtually faultless.

106 l. 212. *idea or fore-conceit*. Sidney uses Plato's word 'idea' and his own phrase 'fore-conceit' to denote the original conception in the artist's mind to which he attempts to give expression in poetry.

l. 215. *imaginative*. Fanciful, with derogatory connotations.

l. 217. *a Cyrus*. Another reference to Xenophon's *Cyropaedia*; see above, l. 220. As the archetypal *Bildungsroman*, Xenophon's work was regularly used in the education of young princes, such as Edward VI and James VI of Scotland, as well as forming part of the curriculum of Shrewsbury School when Sidney was there (T. W. Baldwin, *Small Latine and Lesse Greeke* (1944), 237, 543, 391, and *passim*).

107 l. 241. *a speaking picture*. The saying, much quoted in the Renaissance, that poetry is a speaking picture and painting a silent poetry, was attributed by Plutarch ('*De gloria Atheniensum*', *Moralia*, 346) to the poet Simonides of Ceos.

l. 248. *Emanuel Tremellius and Franciscus Junius*. Tremellius and Junius called these books poetical in their joint translation of the Bible into Latin (Frankfurt, 1575, ii. 4). When previously published by Henri Estienne in 1569 Tremellius's *New Testament* had been dedicated to Queen Elizabeth.

l. 253. *St James's counsel*. James 5: 13: 'Is any among you afflicted? let him pray. Is any merry? let him sing psalms.'

ll. 258–60. *Tyrtaeus . . . Lucan*. Tyrtaeus (cf. ll. 69–71 note) praised martial valour; Phocylides wrote verse precepts: 'Cato' at this time denoted the moral distichs gathered up by Erasmus, much used in the first and second form at grammar schools; Lucretius' *De Rerum Natura* treats of the nature of the physical world, and Virgil's *Georgics* of farming; Manilius was the author of *Astronomica*, a poem in five books; Joannes Jovius Pontanus, the only post-classical author in this list, wrote *Urania*, a neo-Latin astronomical poem, in the late fifteenth century; and Lucan's *Pharsalia*, much admired by the Elizabethans, described the wars between Caesar and Pompey.

l. 266. *this question*. The question of how effectively they teach virtue.

l. 271. *the constant though lamenting look of Lucretia*. Depictions of Lucretia's suicide were extremely common in Renaissance Europe, and found their way into the houses of many Elizabethan noblemen.

ll. 279–80. *waited on . . . poets*. That is, the best writers in the best languages call these writers 'poets', not 'prophets' or 'seers' ('vates').

108 ll. 288–9. *the heroic, lyric, tragic, comic, satiric, iambic, elegiac, pastoral*. This list of eight poetic genres, in what appears to be descending order of magnitude, derives from Horace, Quintilian, and others, but

is unusual in assigning the second highest place to the 'lyric'. The 'iambic' was a form of satire written in iambic metre associated with the early Greek writers Archilochus and Hipponax.

108 l. 292. *numbrous*. Rhythmical, measured; probably a coinage of Sidney's.

l. 299. *as Cicero saith of him*. In his *Epistola ad Quintus*, 1.i.23, Cicero praised Xenophon's *Cyropaedia* as an exemplary fiction, not a history.

ll. 300–1. *Heliodorus . . . Chariclea*. Heliodorus' *Aethiopica*; see above, ll. 206–9.

109 l. 334. *the astronomer, looking to the stars, might fall in a ditch*. Plato, *Theaetetus*, 174A, described the astronomer Thales falling into a well; cf. *AS* 19.10.

l. 340. *architektoniké*. The master-art or science which prescribes to all beneath it' (Liddell and Scott, *Greek Lexicon*).

l. 352. *the moral philosophers*. This account of the futility of philosophy owes much to Cornelius Agrippa, *De incertitudine et vanitate Scientiarum et Artium* (1530), which had been translated into English by James Sanford in 1569.

110 ll. 379–80. *I am testis temporum, lux veritatis, vita memoriae, magistra vitae, nuntia temporis*. Misquotation of a well-known passage in Cicero, *De Oratore*, II. ix. History claims to be 'the witness of the ages, the light of truth, the life of memory, the governess of life, the herald of antiquity' cf. North's *Plutarch*, i. 16).

ll. 383–4. *in the battles of Marathon, Pharsalia, Poitiers, and Agincourt*. The Athenians defeated the Persians at Marathon in 490 BC; Caesar defeated Pompey at Pharsalia in 48 BC; Edward, the Black Prince, captured the King of France at Poitiers in 1356; and Henry V defeated the French at Agincourt in 1415.

l. 392. *Brutus, Alphonsus of Aragon*. According to Plutarch, Marcus Brutus spent his time before the battle of Pharsalia studying history: 'when others slept, or thought what would happen the morrow after, he fell to his book, and wrote all day long, till night, writing a breviary of Polybius'; and Alphonsus of Aragon was said to have cured himself of sickness by reading about Alexander the Great (North's *Plutarch*, vi. 185; i. 17–18).

l. 396. *moderator*. Judge, arbitrator.

111 ll. 406–7. *rather formidine poenae than virtutis amore*. Through fear of punishment rather than love of virtue; cf. Horace, *Epistles*, I. xvi. 52–3.

l. 419. *halt*. Limp, proceed defectively.

ll. 437–8. *a man that had never seen an elephant or a rhinoceros*. This included most Elizabethans; however, Dürer's woodcut (1515) gave sixteenth-century readers an idea of the rhinoceros, and Sidney would have been familiar with *imprese*, copied by Abraham Fraunce from Paolo Giovio, showing an elephant and a rhinoceros (K. Duncan-Jones, 'Two Elizabethan Versions of Giovio's treatise on *imprese*', *English Studies*, 55 (1971), 120–1).

111 l. 440. *the architecture.* The structure (cf. *OED* senses 3, 5).

112 ll. 452–3. *Tully . . . the force love of our country hath in us.* A common theme in Cicero; cf. *De officiis*, I. xxiv. 83–4; xlv. 159–60; III. xxiv. 93; xxv. 95; xxvii. 100; *De oratore*, I. 196–7; *De finibus*, III. 64.

l. 454. *old Anchises . . . Troy's flames.* Virgil, *Aeneid*, II. 634–50.

ll. 455–6. *Ulysses . . . Ithaca.* Homer, *Odyssey*, V. 149 ff., 215.

l. 424. *Anger . . . a short madness. 'Ira furor brevis est'*, Horace, *Epistles*, I. ii. 62.

l. 457. *let but Sophocles bring you Ajax on a stage.* Actually, this is reported, not shown; Sophocles, *Ajax*, 1061. Sidney probably knew the play from J. C. Scaliger's translation of it into Latin.

l. 461. *the schoolmen.* The scholastic philosophers of the Middle Ages; academics who taught in 'schools', or university lecture halls.

ll. 462–3. *Ulysses . . . Euryalus.* All examples from the history of Troy, but not precise enough to be identified with particular passages in Homer, Virgil, or later writers.

ll. 465–7. *Oedipus . . . Agamemnon . . . Atreus . . . the two Theban brothers . . . Medea.* These are all subjects of Greek Tragedy, though Sidney and his readers may have been more familiar with them from the tragedies of Seneca, which have been translated into English between 1559 and 1567.

l. 468. *the Terentian Gnatho.* From Terence, *Eunuchus*; the word 'gnatho' was used to mean 'sycophant, parasite' during the sixteenth and seventeenth centuries.

l. 468. *our Chaucer's Pandar.* The word 'pandar', for a go-between, derives from the character in Chaucer's *Troilus and Criseyde.*

ll. 476–7. *Sir Thomas More's Utopia.* More's *Utopia* (1516) had been translated into English by Ralph Robinson in 1551. Sidney's reservations about the work may derive—as suggested by the phrase 'it was the fault of the man'—from More's political and religious standpoint. However, a distant kinsman, George More, travelled with Sidney on his 1577 embassy, and was the occasion of a eulogy of Sir Thomas More to which Sidney was a witness at a social gathering at Nuremberg (cf. K. J. Höltgen, 'Why are there no wolves in England? Philip Camerarius and a German version of Sidney's table-talk', *Anglia*, 99 (1981), 60–82).

l. 479. *absolute.* Perfect, complete.

ll. 485–6. *Mediocribus . . . columnae.* Horace, *De arte poetica*, 372–3, translated by Ben Jonson as

> But neither men, nor gods, nor pillars meant
> Poets should ever be indifferent.

'That is, mediocrity in poets is rejected by all, including their booksellers (whose wares were displayed in Rome around the columns of buildings)' (Shepherd, 174).

113 l. 491. *Dives and Lazarus.* Luke 16: 19–31.

ll. 492–3. *the lost child and the gracious father.* The parable of the Prodigal Son, Luke 15: 11–32.

l. 512. *Aristotle . . . in his discourse of poesy. Poetics*, IX. 1451b. The Greek terms which follow are accurately glossed by Sidney.

l. 525. *Vespasian's picture right as he was.* Vespasian, as described in Suetonius' *Vita Vespasiani*, xx, was coarse-looking and ugly. Sidney could have seen images of him both in England and on the Continent (cf. K. Duncan-Jones, 'Sidney and Titian', in *English Renaissance Studies: Presented to Dame Helen Gardner* (1980), 9–10).

114 ll. 528–9. *the feigned Cyrus in Xenophon . . . the true Cyrus in Justin.* Xenophon's *Cyropaedia* (translated into English by W. Barker, 1560) was a fictionalized account (see above, ll. 298–9); Justin's *Histories*, a compilation made from the earlier Greek history of Trogus Pompeius, had been translated by Arthur Golding (1564).

l. 530. *the right Aeneas in Dares Phrygius.* The account of the Trojan War by 'Dares Phrygius' was believed by medieval writers, such as Chaucer, to be more authentic than Homer or Virgil; if Sidney had doubts about its authenticity, it would not serve his purpose to air them here.

ll. 533–4. *Canidia . . . was full ill-favoured.* Horace, *Epodes*, V, described the ugly witch Canidia trying to regain her beauty; cf. also *Satires*, I. viii; *Epodes*, III. xvii.

ll. 539–40. *Alexander or Scipio himself.* Quintus Curtius wrote a life of Alexander; Livy in his *Histories*, xxi–xxxii wrote about the mainly admirable, but latterly faulty, Scipio Africanus. Cf. also North's *Plutarch*, iv. 298–386 and vi. 395–431.

l. 543. *doctrine.* Learning, knowledge.

l. 544. *the history.* The historian; cf. *AS* 32.3.

l. 548. *a gross conceit.* 'An undiscriminating understanding' (Shepherd).

ll. 559–65. *Herodotus and Justin . . . Darius.* Herodotus, *History*, III. 153–60; Justin, *Histories*, I.x.

l. 566. *Livy . . . his son.* Livy, *Histories*, I. iii–iv.

ll. 567–8. *Xenophon . . . Cyrus' behalf. Cyropaedia*, VI. i. 39 has such an anecdote of Araspas, whom Sidney has confused with Abradatas.

115 l. 578. *from Dante's heaven to his hell.* This is one of the earliest references to Dante in English literature; see also below, l. 1646.

ll. 586–7. *Ulysses in a storm, and in other hard plights. Odyssey*, V and *passim.*

l. 591. *the tragedy writer.* Euripides, as described by Plutarch. '*Quomodo adolescens poetas audire debeat' (Moralia,* 19); the example was of Ixion, who ended up 'manacled' to a wheel.

ll. 595–6. *see we not valiant Miltiades rot in his fetters?* Miltiades, who had won the battle of Marathon for the Greeks, was imprisoned by his own people; cf. Cicero, *Republic,* I. iii. 5.

ll. 596–7. *The just Phocion and the accomplished Socrates put to death like traitors?* Cf. North's *Plutarch*, v. 108, where the execution of Phocion by the Athenians is compared with that of Socrates.

ll. 597–8. *The cruel Severus live prosperously?* The Emperor Lucius Septimus Severus (d. AD 211).

l. 598. *The excellent Severus miserably murdered?* 'Alexander Severus, Emperor AD 222–35, murdered in his thirtieth year by mutineers' (Shepherd).

l. 599. *Sulla and Marius dying in their beds?* 'Lucius Sulla (138–78 BC), dictator of Rome, whose bitter struggles with Caius Marius (157–86 BC) and the Marian party filled all Italy with strife and terror for twenty years' (Shepherd).

ll. 599–600. *Pompey and Cicero . . . a happiness?* Pompey fled to Egypt after his defeat at Pharsalia, and was killed there (48 BC); Cicero tried to escape from Rome after being proscribed by Mark Antony in 43 BC, and was put to death.

ll. 600–2. *See we not . . . highest honour?* Cato, who continued to resist Caesar after his victory at Pharsalia, was finally driven to suicide at Utica, in Africa; Caesar, the rebel, has given his name to future monarchs (cf. 'Kaiser', 'Tsar'). The Holy Roman Emperor was referred to in Latin as 'Caesar', Sidney's 1577 embassy being '*ad Caesarem*'.

l. 603. *Caesar's own words.* 'He was ignorant of letters', reported by Suetonius, *Julius Caesar,* 77, and extracted by Erasmus as an apothegm.

ll. 609–11. *Cypselus, Periander Phalaris, Dionysius . . . usurpation.* Cypselus and Periander were long-reigning tyrants of Corinth in the 7th century BC; Phalaris was a tyrant of Sicily in the 6th century BC; and Dionysius was tyrant of Syracuse in the 4th century BC.

116 l. 621. *philophilosophos.* A lover of the philosophers.

l. 628. *not gnosis but praxis.* Not knowing but doing.

l. 643. *natural conceit.* Innate reason, inclination.

l. 645. *hoc opus, hic labor est.* Virgil, *Aeneid*, VI.129, 'This is the work, this is the toil' (i.e. 'this is the really difficult bit')—words applied by the Sybil to the journey up from the underworld.

l. 646. *human*: secular, not sacred; 'humane learning'.

117 l. 662. *aloes or rhubarbum.* Two commonly used bitter tasting purgatives.

ll. 671–3. *as Aristotle saith . . . delightful. Poetics*, IV. 1448b; cf. *AS* 34.4.

l. 674. *Amadis de Gaule.* A romance, originally Spanish, later translated into French and extended by Herberay and others, which was one of the sources of *OA*.

l. 677. *Aeneas carrying old Anchises on his back. Aeneid*, II. 705–84.

ll. 679–82. *Whom doth not . . . miserum est.* Turnus, betrothed to Lavinia, is destined to be defeated and supplanted by Aeneas; his tale is told in the later books of the *Aeneid*, and the quotation from XII. 645–6, translated thus by T. Twyne (1584): 'And shall this ground fainthearted dastard Turnus flying view?|Is it so vile a thing to die?'

ll. 686–8. *Plato and Boethius . . . poesy.* Plato's use of poetry has already been cited (ll. 76–86); Boethius, in *De consolatione philosophiae*, personified Philosophy as a female figure, and alternated prose with verse.

l. 689. *indulgere genio.* Persius, *Satires*, V. 151, 'to follow one's natural inclination'.

118 l. 698. *Menenius Agrippa.* This story, told by many Roman historians, is now best known from Shakespeare's adaptation of it in *Coriolanus*, I. i. 95 ff., which may owe something to Sidney's version.

ll. 703–5. *if they were Platonic . . . conceived.* Plato's Academy was said to have 'Let no man enter who is not a geometrician' written over the door.

l. 713. *only words.* Words alone.

ll. 715–16. *Nathan the prophet.* 2 Sam. 12: 1–15.

ll. 723–4. *that heavenly psalm of mercy.* Ps. 51.

l. 729. *end.* Aim, objective.

119 l. 742. *Sannazaro and Boethius.* Sannazaro's *Arcadia* alternated passage of narrative with poems; so did Boethius' *De consolatione* (see above, ll. 686–8).

ll. 742–3. *Some have mingled matters heroical and pastoral.* As Sidney himself did, most notably in *NA*; Montemayor and Ariosto were among those who had previously done so.

l. 750. *Meliboeus.* The name of the speaker in Virgil, *Eclogues*, I, who laments his dispossession. Thomas Watson was to use the name for

Sidney's father-in-law, Sir Francis Walsingham, in his elegy *Meliboeus* (1590).

119 l. 752. *Tityrus.* Meliboeus' interlocutor in *Eclogues*, I, who rejoices in the security afforded him by the reign of Augustus.

ll. 753–5. *sometimes . . . patience.* Cf. Sidney's own beast fable, *OA* 66; Spenser's May Eclogue in *The Shepheardes Calender* and *Mother Hubberds Tale.*

ll. 755–6. *contentions . . . trifling victory.* Exemplified in *OA* 29 (modelled on Virgil, *Eclogues*, III) where the singing competition of Nico and Pas has a dog and cat as prizes.

l. 758. *cock of this world's dunghill.* 'Every cock is proud on his own dunghill' (Tilley, C486).

ll. 760–1. *Haec memini . . . nobis.* Virgil, *Eclogues*, VII. 69–70: 'This I remember, and how Thyrsis, vanquished, strove in vain. From that day it is Corydon, Corydon with us' (i.e. Corydon is the pre-eminent poet).

l. 764. *Heraclitus.* The early Greek philosopher who was said to weep at human folly; cf. *OA* 11.12.

l. 768. *Iambic.* Abusive poem, or lampoon, written in iambic metre, as distinct from the more elaborate and indirect 'satire'.

120 l. 771. *Omne vafer vitium ridenti tangit amico.* From Persius, *Satires*, I: 'the rascal probes every fault of his friend'.

l. 775. *circum praecordia ludit.* From the following line in Persius: 'he plays with the secrets of the heart'.

l. 778. *Est Ulubris, animus si nos non deficit aequuus?* Horace, *Epistles*, I. xi. 30: '[Contentment] is at Ulubrae, if there fail you not a mind well balanced' (Loeb trs.). Ulubrae was a notoriously unpleasant, marshy provincial town.

ll. 779–80. *naughty play-makers . . . odious.* Perhaps a gesture of assent to the thrust of Gosson's *Schoole of abuse*, as the word 'abuse' in the next sentence may indicate.

ll. 790–2. *of a niggardly Demea, of a crafty Davus, of a flattering Gnatho, of a vainglorious Thraso.* Characters from the comedies of Terence: Demea from *Adelphi*, Davus from *Andria*, Gnatho and Thraso from *Eunuchus.*

ll. 793–4. *the signifying badge . . . comedian.* With an art akin to caricature, the comic dramatist makes the dominant vices of these characters vividly apparent; a theory closely approaching Jonson's 'comedy of humours'.

l. 797. *in pistrinum.* 'At the mill', working like a slave.

120 l. 805. *tissue.* Rich fabrics.

ll. 806–7. *the affects of admiration and commiseration.* Feelings of terror and pity, as in Aristotle's *Poetics*, 6.

121 ll. 810–11. *Qui sceptra ... in auctorem redit.* 'The tyrant who rules harshly fears those who fear him; terror returns to its agent'; Seneca, *Oedipus*, III. 705–6; cf. adaptation in *CS* 14.

l. 813. *Alexander Pheraeus.* He 'went out of the theatre ... because he was ashamed his people should see him weep, to see the miseries of Hecuba and Andromache played, and that they never saw him pity the death of any one man, of so many citizens as he had caused to be slain' (North's *Plutarch*, ii. 323).

l. 823. *Lyric.* For the purposes of his argument Sidney adopts here the limited, classical definition of 'lyric' as 'panegyric', hymn of praise.

l. 828. *the old song of Percy and Douglas.* The ballad of *Chevy Chase*, describing the fatal conflict between the Earls of Percy and Douglas.

l. 830. *crowder.* Fiddler. Several documents show Sidney's fondness for popular musicians: for instance, at the age of eleven he gave money to a 'blind harper' (Wallace, 421).

l. 833. *Pindar.* The Greek lyric poet who wrote hymns and accounts of military and sporting triumphs; his notoriously elaborate metres contrast with the simple four-line stanza of the Border ballad.

l. 833. *In Hungary.* Sidney was in Hungary for a few weeks in the late summer of 1573 (Osborn, 102–4). The Hungarians' 'soldierlike' powers were frequently tested in battles with the invading Turks.

ll. 836–7. *The incomparable Lacedaemonians.* The Lacedaemonians, or Spartans, were much admired by the Elizabethans. Plutarch described their warlike music in the *Life of Lycurgus* (North's *Plutarch*, i. 148–50).

ll. 846–7. *Philip of Macedon ... three fearful felicities.* Cf. North's *Plutarch*, iv. 300; his other 'felicities' were a victory over the Illyrians and the birth of Alexander, all three occurring on the same day.

122 l. 854. *Tydeus.* One of the 'seven against Thebes' in Statius, *Thebais*.

l. 854. *Rinaldo.* Hero of Tasso's *Gerusalemme Liberata.*

ll. 857–8. *the saying of Plato and Tully.* Plato, *Phaedrus*, 250D; Cicero, *De finibus*, II. xvi. 52 and *De officiis*, I. v. 14; and cf. *AS* 25. 1–4.

l. 880. *melius Chrysippo et Crantore.* Horace, *Epistles*, I. ii. 'better than Chrysippus and Crantor' (two early Greek philosophers).

123 ll. 892–3. *prophesying ... making. 'Vates'*, 'poet'.

ll. 904–5. *even our Saviour Christ vouchsafed to use the flowers of it.* In parables, like that of the Prodigal Son; see above, ll. 566–8.

l. 919. *the spleen.* Thought to be the seat of hostile laughter.

123 ll. 924–5. *a playing wit . . . plague.* Three examples of the mock encomium, a favourite humanist genre with which *DP* itself has some kinship. Lucian, Apuleius, and Cornelius Agrippa were among those who eulogized the ass; Francesco Berni (*c.* 1496–1535) wrote comic poems in praise of debt and plague (*Il primo libro dell'opere burlesche* (Florence, 1558), 9–18, 80–7).

124 l. 928. *Ut lateat virtus proximitate mali.* Adapted from Ovid, *Ars Amatoria,* II. 662 (Sidney translates the phrase).

ll. 929–31. *Agrippa . . . folly.* Cornelius Agrippa, *De incertitudine et vanitate scientiarum et artium* (1530); Erasmus, *Moriae Encomium* (1511).

ll. 932–3. *for Erasmus . . . promise.* These apparently sportful works were more profound and serious than might be thought (perhaps like *DP*).

ll. 936–7. *scoffing cometh not of wisdom.* Sounds proverbial, but not traced.

l. 944. *Scaliger judgeth. Poetices,* I. ii.

ll. 945–6. *oratio . . . mortality.* Commonplace idea deriving from Cicero and other classical writers that reason and language distinguish man from beasts.

l. 950. *without.* Unless.

ll. 954–5. *memory being the only treasure of knowledge.* Tilley, M870.

ll. 964–6. *they that have taught the art of memory . . . known.* Formal theories of the 'art of memory', propounded by Raymond Lull and others, recommended the identification of sections of argument with objects which were then allocated to particular positions in an imaginary building (cf. F. A. Yates, *The Art of Memory* (1966)).

125 ll. 985–6. *the largest field to ear, as Chaucer saith. Knight's Tale,* 28.

ll. 989–90. *as if they had overshot Robin Hood.* Extravagantly boastfully; cf. Tilley, R 148: 'Many speak of Robin Hood that never shot in his bow'.

ll. 990–1. *Plato banished them out of his commonwealth. Republic,* II. iii. 10.

l. 993. *petere principium.* 'Beg the question'.

126 l. 1009. *Charon.* The ferryman who rowed dead souls over the River Styx to the underworld.

l. 1013. *artists.* Men skilled in 'liberal arts'.

l. 1016. *maketh any circles.* Refers to necromancy, in which spirits were summoned into a circle.

l. 1040. *John-a-stiles and John-a-nokes.* John of the stile and John of the oaks; fictitious names used in teaching or debating legal arguments.

127 ll. 1050–1. *the principal . . . abuse I can hear alleged.* The stimulus given by drama—especially stage comedy—to sexual licence was the dominant theme of Gosson's *Schoole of abuse* (1579), though a wider reference may be intended, since the other genres are not much dealt with by Gosson.

l. 1054. *even to the heroical Cupid hath ambitiously climbed.* For instance, in the mingled epics of Ariosto and Tasso; also Sidney's own *OA*.

ll. 1061–2. *Some of my masters . . . the excellency of it.* Perhaps a dig at Plato, among others, who eulogized love in the *Symposium* and the *Phaedrus.*

ll. 1070–1. *eikastiké . . . phantastiké.* 'showing forth', or 'picturing'; 'fanciful, imaginative'.

128 ll. 1102–3. *never was the Albion nation without poetry.* In his *Description of Britain* William Harrison gave an account of Britain as a place where the arts flourished even before it was conquered by 'Albion' (Holinshed, *Chronicles* (1577), 1–2).

l. 1104. *chainshot.* Cannon balls linked by a chain; hence, a resounding blow.

l. 1106. *certain Goths.* Continuation of Dio Cassius, *Roman Histories,* LIV. 17.

l. 1107. *hangman.* Villain, rogue.

l. 1119. *jubeo stultum esse libenter.* Adapted from Horace. *Satires,* I. i. 53, 'I willingly tell him to be a fool'. Quoted in this form by Francis Davison in the Preface to *A poetical Rapsody* (1602).

l. 1123. *the quiddity of ens and prima materia.* Terms from scholastic philosophy: 'the essential nature of being and original matter.'

l. 1123. *a corslet.* Body armour.

129 ll. 1133–4. *Alexander . . . took dead Homer with him.* Alexander loved the *Iliad* so much that he 'laid it every night under his bed's head with his dagger' (North's *Plutarch*, iv. 305; cf. also Plutarch's essay *De Fortuna aut virtute Alexandri*).

l. 1143. *the former.* Cato the Censor (234–149 BC), rather than his great grandson, Cato of Utica (see above, l. 601); cf. North's *Plutarch*, iii. 1–47. The story of Cato's dislike of Fulvius' bringing the poet Ennius on campaign with him was often repeated, e.g. by Cornelius Agrippa and Stephen Gosson.

l. 1149. *unmustered.* Unenrolled (as a soldier).

ll. 1152–4. *the other Scipio brothers . . . sepulture.* Cicero, *Pro Archia poeta,* ix. 22, mentions that the poet Ennius was buried in the Scipios' vault.

130 ll. 1172–4. *they found for Homer . . . to live among them.* Cicero, *Pro Archia poeta,* viii. 19, is among those who said that seven Greek cities competed for the honour of having Homer among them. Empedocles and Protagoras were among Greek philosophers banished from their native cities.

ll. 1175–7. *For only repeating . . . unworthy to live.* The story of some Athenians being saved from slaughter by the Syracusans for reciting lines from Euripides is told by Plutarch in his *Life of Nicias* (North's *Plutarch,* iv. 42–3); there is an allusion in the same work to the banishment of the philosopher Protagoras and the execution of the philosopher Socrates (ibid. 35).

ll. 1177–9. *Certain poets . . . a just king.* Hieron I, tyrant of Syracuse, was a patron of art and literature. Simonides reconciled Hieron to his brother Theron, and Pindar celebrated Hieron's achievements.

ll. 1179–80. *where Plato . . . was made a slave.* Cf. Cicero, *Pro Rabirio Postumo,* ix. 23.

ll. 1184–5. *see whether any poet do authorize abominable filthiness as they do.* Like Scaliger (*Poetices,* I. ii) Sidney believes Plato to authorize homosexuality; and in Plutarch's *De amore (Moralia,* 751) Protogenes asserts that 'there is only one genuine Love, the love of boys'.

ll. 1187–8. *where he himself alloweth community of women. Republic,* v.

ll. 1193–4. *St Paul . . . 'their prophet'.* In the Penshurst MS of *DP* this citation is glossed: 'Acts: 17. To Titus: 1'.

ll. 1205–6. *Plutarch . . . divine providence. De Iside et Osiride, Moralia,* 351–84; *De defectu oraculorum, Moralia,* 410–38; and probably *De sera numinis vindicta, Moralia,* 548–68.

131 ll. 1213–14. *Qua authoritate . . . exigendos.* Scaliger, *Poetices,* I. ii, refers to Plato 'whose authority certain barbarous and uncouth men seek to use to banish poets from the commonwealth'.

ll. 1218–20. *who . . . unto poetry.* 'Whatever Plato intended, the dialogue [*Ion*] was never taken ironically by 16th century readers, who all found in it many passages commending poetry' (Shepherd). One such commentator was Landino; see below, l. 1644.

ll. 1223–4. *under whose lion's skin . . . poesy.* Refers to Aesop's fable of the ass who put on a lion's skin and passed for a lion until a stranger who had seen real lions unmasked him (Aesop, *Fables,* 120–1).

l. 1227. *more than myself do.* Sidney was unusual among Renaissance theorists in rejecting the idea of divine inspiration for poetry (cf. *AS* 74.4–5), but conventional in taking Plato's *Ion* to support the notion.

ll. 1233–5. *Laelius . . . made by him.* Terence hints in the prologue to the *Heautontimoramenos* ('The man who hurts himself') that parts of it were written by Gaius Laelius, friend of Scipio Africanus the younger.

131 ll. 1235–7. *even the Greek Socrates . . . verses.* Plato, *Phaedo,* 60, Plutarch, *Quomodo adolescens poetas audire debet, Moralia,* 16. For Socrates as 'the only wise man', cf. *AS* 25.1–2.

l. 1243. *guards.* Borders, decorative trimmings.

132 l. 1258. *career.* Race course, or running track in the tiltyard; Sidney is imagining himself either on horseback, or as himself a horse.

l. 1264. *Musa . . . laeso?* Virgil, *Aeneid,* 1.8, 'Tell me, O muse, the cause, wherein thwarted in will . . . ?'

ll. 1266–7. *David . . . to be poets.* David was both King of Israel and author of the Psalms; Hadrian was Emperor of Rome (AD 117–38) and wrote verses; Sophocles was an Athenian general as well as a dramatist; Germanicus, nephew of Tiberius, conquered Germany and was reputed to write poetry.

ll. 1268–9. *Robert . . . King James of Scotland.* Robert II of Anjou (1309–43) was a patron of Petrarch; Francis I of France was patron of many artists and writers, including Rabelais, Erasmus, and Leonardo da Vinci; James I of Scotland (1394–1437) was a patron of letters and author of *The Kingis Quhair.* James VI, future James I of England, was born in 1566, so cannot have been much more than 16 when Sidney wrote *DP.* However, Sidney alludes to his tutor, Buchanan, further on in the sentence, and may have been aware of James's youthful verses which were to be published as *Essayes of a Prentis* (1584). Shepherd (212) and van Dorsten (59) suggest that the allusion is to James VI, but that it breaks the pattern of pairs in Sidney's list, and may be a later insertion. However, the list proceeds in a fairly orderly fashion, with steady diminution in the number of examples, thus: 4; 3; 2, 2, 2, 2; 1, 1; the overall organization of the sentence does not support the idea of revision and insertion. James VI's contribution of an English sonnet to the Cambridge volume of elegies on Sidney (*Academiae Cantabrigiensis lachrymae* (1587), sig. K1) could reflect some earlier mutual knowledge of literary endeavours. The present allusion may or may not be to him.

l. 1270. *Bembus and Bibbiena.* Pietro Bembo (1470–1547) wrote poems and prose, including a defence of the Tuscan language analogous to Sidney's own defence of English in the succeeding pages; Bernard Dovizi, Cardinal Bibbiena (1470–1520), was secretary to Lorenzo de'Medici and wrote a Plautine comedy.

l. 1271. *Beza and Melanchthon.* Theodore de Bèze (1519–1605) wrote a tragedy, *Abraham sacrifiant*, and an edition of the New Testament. His commentary on the Psalms was drawn on by Sidney in his translation. Philip Melanchthon (1497–1560) was a German humanist poet and educationalist, close associate both of Luther and of Sidney's friend's friend Hubert Languet.

132 l. 1272. *Fracastorius.* Girolamo Fracastorio (1483–1553), author of medical and scientific works, some in verse.

ll. 1272–3. *Pontanus and Muretus.* Giovanni Pontano (1426–1503), astronomical poet (see above, l. 259); Marc-Antoine Muret (1526–85), French humanist and scholar.

l. 1273. *George Buchanan.* Scottish humanist, tutor to James VI, and translator of the Psalms (cf. I. Macfarlane, *George Buchanan* (1981).

l. 1274. *that Hôpital of France.* Michel Hurault, de l'Hôpital (1503–73), Chancellor of France 1560–8, whose support for religious tolerance must have commended him to Sidney, was a Latin poet, who wrote six books of verse epistles. Both Buchanan and he may have been personally known to Sidney

ll. 1281–3. *heretofore . . . did sound loudest.* Sidney is unspecific about when this time was, perhaps deliberately, since he later (l. 1339) refers to Chaucer's period as a 'misty time'.

l. 1284. *strew the house.* Prepare a welcome by spreading fresh rushes and flowers on the floor. Sidney did not regard the peace enjoyed by England in Elizabeth's reign as an unmixed blessing.

l. 1285. *the mountebanks at Venice.* Quick-tongued salesmen, 'mounted on a bench', who made a strong impression on English visitors to Venice such as Thomas Coryate. Jonson's Volpone assumes the role of a mountebank selling quack medicine, *Volpone,* II. ii.

l. 1291. *base men with servile wits.* As in *AS* 3 and 15, Sidney implies that there are a lot of bad, derivative poets among his contemporaries, but names no names.

ll. 1293–5. *Epaminondas . . . highly respected.* Plutarch, in his *Precepta gerendae reipublicae (Moralia,* 811) describes Epaminondas accepting the office of telearch (which involved organizing street-cleaning) and bringing it distinction (cf. D. A. Russell, *Plutarch* (1973), 14).

133 l. 1298. *without any commission.* i.e. no one has asked them to write poetry.

l. 1301. *Queis meliore luto finxit praecordia Titan.* Juvenal, *Satires,* XIV. 33–5, 'whose souls informing better clay Prometheus has shaped'.

l. 1303. *knights of the same order.* A jocular or contemptuous phrase; cf. *OED* 12c, 'knight of the pen'.

l. 1320. *orator fit, poeta nascitur.* Tilley, P 451, 'Poets are born but orators are made'. Keats adapts the same idea in saying 'That if Poetry come not as naturally as the Leaves to a tree it had better not come at all'. (*Letters*, ed. M. Buxton Forman (1952), 107).

133 l. 1322. *a Daedalus.* In Greek mythology, the originator of the arts, father of the high-flying Icarus whose wings melted in the sun; used here allegorically for 'the foundation of poetry'.

l. 1326. *we.* We 'paper-blurrers'.

l. 1327. *fore-backwardly.* Back to front (a coinage of Sidney's).

l. 1332. *quodlibet.* 'Anything we like'; any question proposed in scholastic debate.

134 l. 1334. *Quicquid conabor dicere, versus erit.* Ovid, *Tristia*, IV. x. 26, 'whatever I tried to say turned out as poetry'. (Cf. Pope, *Epistle to Dr Arbuthnot*, 128, 'I lisp'd in Numbers, for the Numbers came'.)

ll. 1337–40. *Chaucer . . . after him.* The five-book structure of *Troilus and Criseyde*, with the lovers' consummation described in the central book, may have been Sidney's prime model for the structure of *OA*.

ll. 1341–2. *I account the Mirror of Magistrates . . . beautiful parts. The Mirror for Magistrates* was a cumulative collection of poems, initiated by Thomas Sackville in 1559, on the falls of kings and eminent people. Poems on ancient British monarchs (such as Cordelia) had been added in the 1574 edition. It is a work made up of 'parts', not a coherent whole (cf. edition by L. B. Campbell, 2 vols., 1938, 1946).

ll. 1342–4. *the Earl of Surrey's lyrics . . . noble mind.* A substantial proportion of Surrey's lyrics were included in *Tottel's Miscellany* (1557). Some contemporaries saw Sidney as Surrey's poetic heir; cf. G. Whitney, *A choice of emblems* (1586), 196–7.

ll. 1344–5. *The Shepheardes Calender . . . deceived.* Spenser's *Shepheardes Calender* (1579) was dedicated to Sidney under the pseudonym 'Immerito'. Perhaps, at the time of writing *DP*, Sidney was unaware that Immerito was Edmund Spenser; but more probably he deliberately respected his pseudonymity. In his dedicatory poem Spenser commends the book to Sidney 'As child whose parent is unkent'. Sidney's criticism of Spenser's language in the succeeding sentence may be proprietorial in tone, rather than dismissive. It is not quite just, since Theocritus did use Doric dialect for many of his *Idylls*: and cf. also *OA* 66. However, the 'marked tendency towards modernization' identified by De Sélincourt in the third edition of the *Calender* (1586) may reflect a response to Sidney's comment (Spenser, *Minor Poems*, ed. E. De Sélincourt (1910), xii).

l. 1346. *allow.* Praise.

ll. 1357–63. *Gorboduc . . . might not remain as an exact model of all tragedies. Gorboduc*, named on the title-page of the first authorized edition (1571) as *The Tragidie of Ferrex and Porrex*, had been performed at Court in 1561 and was the joint work of Thomas Sackville, later Lord Buckhurst (1536–1608) and Thomas Norton (1532–84). This political fable of the evils ensuing to a divided Britain was a model or reference point for many writers of the period, including Shakespeare. Sidney accords it high praise in finding its style comparable with that of Seneca's tragedies, but finds it deficient in 'circumstances', that is, the Aristotelian unities of place and time.

134 l. 1368. *inartificially*. Inartistically.

135 l. 1382. *traverses*. Troubles, difficulties.

ll. 1386–7. *the ordinary players in Italy*. Sidney may have seen such players during 1574, when he was in Venice and Padua (cf. K. M. Lea, *Italian Popular Comedy* (1934), i. 262).

ll. 1387–9. *Yet will some bring in . . . twenty years*. As elsewhere in *DP*, Sidney appears to be writing without checking his citations; Terence's *Eunuchus* occupies a single day, but he may really be thinking of the longer-drawn-out *Heautontimorumenos* (see above, ll. 1234–5) which Scaliger (*Poetices*, VI. iii) described as performed in two parts on two successive days.

ll. 1390–1. *Plautus have in one place done amiss*. Probably another recollection of Scaliger (ibid.), who criticizes the extended action of Plautus' *Captives*.

ll. 1400–1. *Peru . . . Calicut*. Sidney imagines remote and recently discovered points West and East: Peru, conquered by Pizarro (1478–1541) and Calicut (Calcutta) discovered by Vasco da Gama in 1498.

l. 1402. *Pacolet's horse*. The horse belonging to the enchanter Pacolet in the popular late medieval romance *Valentine and Orson*.

l. 1403. *Nuntius*. Messenger.

l. 1405. *ab ovo*. Horace, *De arte poetica*, 147, 'from the egg'.

136 l. 1418. *Euripides*. In his *Hecuba*, from which there are concealed quotations in the phrases 'for safety's sake, with great riches' and 'to make the treasure his own' (cf. K. O. Myrick, *Sir Philip Sidney as a Literary Craftsman* (1935; repr. 1965), 105–7).

l. 1427. *Apuleius did somewhat so*. In his *Golden Ass*, translated by W. Adlington (1566), a source for *OA*; this is not a drama, however, so the example is an odd one.

l. 1430. *Plautus hath Amphitryo*. Scaliger and others regarded this play as the archetypal tragi-comedy.

l. 1431. *daintily*. Reluctantly.

ll. 1445–6. *Delight . . . Laughter hath only a scornful tickling*. Sidney's account of comedy is a modified version of the theories of Castelvetro and Trissino; cf. Shepherd, 223–5.

136 l. 1454. *go down the hill against the bias.* Travel down a slope which counteracts the direction given to a bowl by its bias.

137 l. 1459. *Alexander's picture well set out.* Alexander was painted by Apelles and others (North's *Plutarch*, iv. 300–1); Signorelli and Giulio Romano are among the many Renaissance artists who depicted him, and there were some pictures of him in Elizabethan houses, e.g. one at Essex House (C. L. Kingsford, 'Essex House', *Archaeologia*, 73 (1923), 1–54).

ll. 1460. *twenty mad antics.* Referring to the 'antique' or grotesque style of decoration, described thus by Peacham: 'an unnatural or unorderly composition for delight sake, of men, beasts, birds, fishes, flowers &c., without (as we say) rhyme or reason; for the greater variety you show in your invention the more you please' (*The Gentleman's Exercise* (1612), 50).

ll. 1460–2. *Hercules . . . spinning at Omphale's commandment.* After the completion of his twelve Labours, Hercules became infatuated with Omphale, Queen of Lydia; Plutarch refers to comic pictures of him in women's clothes (*Moralia,* 785), and similar pictures were fairly common in the Northern European Renaissance, e.g. by Cranach and by Bartholomaeus Spränger, court painter to the Emperor Rudolph.

ll. 1475–6. *Nil habet . . . quod ridiculos homines facit?* Juvenal, *Satires*, III. 152–3, 'Poverty contains no sharper misery than that it makes men ridiculous'. We may note that Sidney's comic clowns, the Dametas family in *OA* and *NA*, have been elevated from poverty to royal favour, and so become legitimate butts of merriment.

l. 1478. *Thraso.* A boaster, like the character in Terence's *Eunuchus.*

l. 1478. *A self-wise-seeming schoolmaster.* Like Rombus in *LM*.

l. 1481. *the other.* In tragedy. As Shepherd observes (225), 'the complimentary reference to George Buchanan as a model for the writer of tragedies has something of the appearance of an afterthought'.

138 ll. 1497–8. *so coldly they apply fiery speeches.* The Petrarchan oxymoron of 'icy fire', used here to condemn third-generation Petrarchizers.

l. 1503. *forcibleness or energia.* The word 'energy' was not yet current in English; '*energeia*' (Greek) or '*efficacia*' (Latin) were used by rhetoricians and literary theorists such as Scaliger to denote 'the force of poetry', or the clarity with which the poetic theme has been realized in words.

ll. 1509–13. *far-fet words . . . winter-starved.* List of poetic faults—affected diction, excessive alliteration, Euphuistic similes—which closely parallels *AS* 15.

138 ll. 1518–21. *the diligent imitators of Tully and Demosthenes . . . make them wholly theirs.* An attack on excessive and excessively overt Ciceronianism. 'Nizolian paper-books' were commonplace books of phrases from the classical rhetoricians, named after Marius Nizolius who published *Thesaurus Ciceronianus* (1535). Sidney's friend Henri Estienne wrote an attack on the 'Nizolian' method, *Nizoliodidascalus* (1578). Sidney seems to recommend, rather, absorption in the original writers and a personal transformation ('translation') of their techniques.

ll. 1525–8. *Tully, when he was to drive out Catiline . . . Imo in senatum venit &c.* From the opening of Cicero's *In Catilinam*, I: '[What an age we live in! The senate knows it all, the consul sees it, and yet] this man is still alive. Alive did I say? Not only is he alive, but he attends the senate' (Loeb trs.). Sidney quotes from the same well-known passage in *LM* (3).

l. 1532. *too too much choler to be choleric.* Excessive rage to express rage with so much repetition. There may be a pun on 'colour', or figure of rhetoric.

l. 1533. *similiter cadences.* From Latin *similiter cadentiae*, similar endings; Sidney refers to the use of rhyme, assonance, and other figures of repetition by public speakers.

l. 1534. *daintiness.* Discretion. Less imitation of Cicero, more of Demosthenes, is what Sidney seems to recommend.

ll. 1535–7. *the sophister . . . had none for his labour.* A commonplace story about the pitfalls of specious logic, told, for instance, by Thomas More: 'as though a sophister would with a fond argument prove unto a simple soul that two eggs were three, because that there is one, and there be twain, and one and twain make three: that simple unlearned man, though he lack learning to foil his fond argument, hath yet wit enough to laugh thereat, and eat two eggs himself, and bid the sophister take and eat the third' ('The Confutation of Tyndale's Answer III', in *Works*, Yale edition, viii. 287).

139 ll. 1540–1. *certain printed discourses.* Perhaps a reference to Lyly's *Euphues, the Anatomy of Wit* (1578) and its sequel, *Euphues and his England* (1580). Sidney's dislike of Lyly's style could have been compounded by the fact that his patron was the Earl of Oxford.

ll. 1550–5. *Antonius and Crassus . . . used these knacks very sparingly.* Cicero, *De Oratore*, II. i.

l. 1558. *curiously.* Elaborately, painstakingly.

l. 1564. *to hide art.* Tilley, A335: 'It is art to hide art'.

l. 1566. *pounded.* Rounded up into a 'pound' or enclosure, like a straying animal; Sidney may be still imagining himself a horse (cf. ll. 21–2).

139 ll. 1567–8. *in the wordish consideration.* In matters of diction and style.

l. 1576. *it is a mingled language.* Unlike Sidney, other humanist critics regretted the mixed character of English, e.g. 'E.K.', in his Preface to Spenser's *Shepheardes Calender*: 'they have made our English tongue a gallimaufray or hodgepodge of all other speeches.'

140 ll. 1581–2. *the Tower of Babylon's curse.* Elizabethans identified 'Babylon' with 'Babel' (Gen. 10: 10).

ll. 1585–8. *is particularly happy . . . can be in a language.* Sidney's inventive use of compound words and epithets was recognized by later writers as one of his most distinctive achievements.

l. 1589. *of versifying there are two sorts.* The dialogue between Dicus and Lalus at the end of two texts of the *OA* First Eclogues (*OA* 89–90) can be seen as an early draft for this passage in *DP*. For a full discussion of the metrical debate and Sidney's role in it, see Derek Attridge, *Well-weighed syllables* (1974).

l. 1601. *vulgar.* Vernacular, not learned.

l. 1602. *the ancient.* The classical, quantitative method of versification.

l. 1603. *the Dutch.* German. Sidney complained of the harshness of the German language in one of his letters to Languet.

l. 1608. *dactyls.* Long (strong) syllable followed by two short (weak) syllables.

l. 1617. *sdrucciola.* 'Slippery' or 'sliding' rhyme, where the rhyme words are trisyllables, used extensively by Sidney, e.g. in *OA* 7. John Florio defines it differently, as 'a kind of smooth running blank verse'. (*Queen Anna's New World of Words* (1611)), but this is not what is meant here.

141 l. 1628. *poet-apes.* False, imitative poets; cf. *AS* 3.3.

l. 1634. *Aristotle.* As cited by Boccaccio, *De genealogia deorum*, XIV. viii.

l. 1636. *Scaliger. Poetices*, III. xix.

ll. 1638–41. *Clauserus . . . quid non?* Conrad Clauser, preface to his translation of L. Annaeus Cornutus, *De natura deorum gentilium* (Basle, 1543).

l. 1644. *Landino*: Cristoforo Landino, prologue to his edition of Dante's *Divina commedia* (Florence, 1481). We may notice that Sidney is here conjuring his readers to believe in the poet's 'divine fury' though he has himself previously rejected the idea (1220).

l. 1651. *libertina patre natus . . . Herculea proles.* The first phrase is from Horace, *Satires*, I. vi. 6, where in celebrating his friendship with his patron Maecenas he calls himself 'the son of a freedman'; the second phrase means 'descendant of Hercules'.

141 l. 1652. *Si quid mea carmina possunt.* Virgil, *Aeneid*, IX. 44, 'If my verses can achieve anything'.

ll. 1654–5. *the dull-making cataract of Nilus.* The deafening effect of the waterfalls in the upper reaches of the Nile was described by Cicero, *Somnium Scipionis*, V. 13; in the same work he writes of the music of the spheres.

142 l. 1658. *such a mome as to be a Momus.* Such a dunce ('mome') as to be a carping critic. Momus, son of Night, was a type of the bad-tempered fault finder.

l. 1659. *the ass's ears of Midas.* Awarded to King Midas because he preferred the music of Pan to that of Apollo. (Ovid, *Metamorphoses*, XI. 146 ff.)

l. 1660. *Bubonax.* Bupalus, who made a statue of the ugly poet Hipponax, was driven to suicide by Hipponax's verses (Pliny, *Natural History*, XXXVI. v. 4). Sidney has conflated the two names.

l. 1661. *to be rhymed to death, as is said to be done in Ireland.* This example is close to home. Sidney's own father was at one stage threatened with death by Irish bards (cf. K. Duncan-Jones, 'Irish Poets and the Sidneys', *English Studies*, 44: 424–5). Cf. also Scot, *Discoverie of Witchcraft* (1584), III. xv, 'The Irishmen . . . will not stick to affirm, that they can rhyme either man or beast to death.'

Further Reading

(This list supplements the titles included in the Abbreviations, 143–5, which are not repeated here.)

BOOKS

Attridge, Derek, *Well-weighed syllables* (1974).
Connell, Dorothy, *Sir Philip Sidney: The maker's Mind* (1977).
Donow, H. S., *A Concordance to the poems of Sir Philip Sidney* (Ithaca, NY, 1975).
Fowler, Alastair, *Conceitful Thought* (1975).
Hamilton, A. C., *Sir Philip Sidney: A Study of his Life and Works* (1977).
Helgerson, Richard, *The Elizabethan Prodigals* (Berkeley, Calif., 1976).
Howell, Roger, *Sir Philip Sidney: The Shepherd Knight* (1968).
Kalstone, David, *Sidney's Poetry: Contexts and Interpretations* (Cambridge, Mass., 1965).
McCoy, Richard, *Sir Philip Sidney: Rebellion in Arcadia* (New Brunswick, 1979).
Myrick, K. O., *Sir Philip Sidney as a Literary Craftsman* (Cambridge, Mass., 1935; repr. 1965).
Nichols, J. G., *The Poetry of Sir Philip Sidney* (1974).
Norbrook, David, *Poetry and Politics in the English Renaissance* (1984).
Patterson, Annabel, *Censorship and Interpretation* (Madison, Wis., 1984).
Pears, S. A., *The Correspondence of Sir Philip Sidney and Hubert Languet* (1845).
Robinson, F. G., *The Shape of Things Known: Sidney's Apology in its Tradition* (Cambridge, Mass., 1972).
Salzman, Paul, *English Prose Fiction 1558–1700: A Critical History* (1985).
Wilson, Mona, *Sir Philip Sidney* (1931, repr. 1950).
Zandvoort, R. W., *Sidney's Arcadia: A Comparison of the two versions* (Amsterdam, 1929).
Zim, Rivkah, *The English Metrical Psalms: Poetry as Praise and Prayer 1535–1601* (1987).

ARTICLES

Bergbusch, Martin, 'Rebellion in the *New Arcadia*', *PQ* 53 (1974), 29–41.
Chaudhuri, S., 'The Eclogues in Sidney's *New Arcadia*', *RES* 35 (1984), 185–202.
Craig, D. H., 'A Hybrid Growth: Sidney's Theory of Poetry in *An Apology for Poetry*', *ELR* 10 (1980), 183–201.
Duncan-Jones, K. D., 'Sidney's Urania', *RES* 17 (1966), 124–32.

Fabry, F. J., 'Sidney's Verse Adaptations to Two Italian Art-songs', *RQ* 23 (1970), 237–55.

Heninger, S. K., 'Sidney and Serranus' Plato', *ELR* 13 (1983), 146–61.

Höltgen, K. J., 'Why are there no wolves in England? Philip Camerarius and a German Version of Sidney's Table-Talk', *Anglia* 99 (1981), 60–82.

Lanham, Richard A., 'Sidney: The Ornament of his Age', *Southern Review* (Adelaide, 1967), 319–40.

Levy, F. J., 'Sir Philip Sidney Reconsidered', *ELR* 2 (1972), 5–18.

Marotti, A. F., '"Love is not love": Elizabethan Sonnet Sequences and the Social Order', *ELH* 49 (1982), 396–428.

Parker, R. W. 'Terentian Structure and Sidney's Original *Arcadia*', *ELR* 2 (1972), 61–78.

Spencer, Theodore, 'The Poetry of Sidney', *ELH* 12 (1945), 251–78.

Warkentin, Germaine, 'Sidney's *Certain Sonnets*: Speculations on the Evolution of the Text', *The Library*, sixth series, 2 (1980), 430–44.

Wilson, Christopher R., '*Astrophil and Stella*: A Tangled Editorial Web', *The Library*, sixth series 1 (1979), 336–46.

Selective Glossary

affects, emotions, affections; cf. Latin *affectus*
appassionate (*adj.*), full of passion
baiting place, place for rest and refreshment (food)
bate (*n.*), dispute, debate
bias, irregularity designed to determine the course of a bowl's path (in game of bowls)
brawl, from French *branle,* a kind of dance
cleeves, cliffs
concent, harmony
contentation, contentment
gins, traps
grateful, pleasing, attractive
haling, pulling, dragging
honesty, chastity
jarl (*vb.*), quarrel
jurat, sworn witness
lickerous, greedy, lecherous
mich (*vb.*), play truant; *micher* (*n.*), one who plays truant
moods, rages.
narr, nearer
niggard (*adj.*), miserly, grudging; *niggardly* (*adv.*), grudgingly
passenger, traveller, passer-by
peise (*vb.*), weigh; *peised* (*adj.*), weighed, balanced
plot, map, plan
prest, prompt, ready
quintessence, the imagined 'fifth element', which alchemists tried to extract
rampire, rampart, defence
rebeck, three-stringed musical instrument
richess, richness
shrewd, mischievous, vicious
spill, destroy
sublime (*vb.*), extract
suitable, matching, appropriate
try, experience, enjoy
waymenting, lamenting
winter-starved, killed or famished by cold
wrack (*n.*), destruction

Index of First Lines

OXFORD POETRY LIBRARY

WILLIAM WORDSWORTH

Edited by Stephen Gill and Duncan Wu

Wordsworth was one of the most illustrious of the Romantic poets. In this selection generous extracts are given from his important work *The Prelude*, together with many of his shorter poems. The reader will find classics such as *Tintern Abbey*, *Westminster Bridge* and 'I wandered lonely as a cloud' well represented. Notes and introduction are provided by Wordsworth's biographers, Stephen Gill and Duncan Wu.

OXFORD POETRY LIBRARY

ALEXANDER POPE

Edited by Pat Rogers

Pope has been acknowledged as the most important poet of the first half of the eighteenth century. This selection includes his brilliant poems *An Essay on Criticism*, *Windsor Forest*, and his masterpiece of social satire, *The Rape of the Lock*. Together with a representative sample of Pope's other verse, Pat Rogers gives an eloquent defence of Pope's poetic practice.

OXFORD POETRY LIBRARY

SAMUEL TAYLOR COLERIDGE

Edited by Heather Jackson

Coleridge was one of the most significant figures in the development of Romantic poetry. This new selection represents the full range of his poetic gifts, from his early polemic poetry such as the *Sonnets on Eminent Characters*, to the maturity of the blank verse poems, *Fears in Solitude* and *Frost at Midnight*. Also included are the wonderful works, *Kubla Khan* and *The Rime of the Ancient Mariner*.

OXFORD POETRY LIBRARY

LORD BYRON

Edited by Jerome J. McGann

Byron was one of the most acclaimed writers of his time, and he continues to be a highly popular Romantic poet with readers today. His mastery of a sweeping range of topics and forms is clearly reflected in this selection, which includes extracts from all his major poems such as *Childe Harold*, *Beppo*, and *Don Juan*, together with many shorter lyrics.

OXFORD POETRY LIBRARY

JOHN DRYDEN

Edited by Keith Walker

Dryden was the leading poet of his day, and dominated the literary scene with satires such as *MacFlecknoe* and *Absalom and Achitophel*. This selection represents the full range of his talent and pays particular attention to his classical translations, which gave new life to English verse. These include extracts from Horace, Lucretius, Ovid, and Virgil's *Aeneid*, as well as his own fables and reworkings of some of Chaucer's tales.

OXFORD POETRY LIBRARY

THOMAS HARDY

Edited by Samuel Hayes

Thomas Hardy continues to be one of the best loved of the great English poets. His enduring popularity is perhaps due to the universality of his subject matter: birth, childhood, love, marriage, age, death. These subjects are well represented in this selection which contains poems taken from all eight of Hardy's poetry volumes.

OXFORD POETRY LIBRARY

ANDREW MARVELL

Edited by Keith Walker and Frank Kermode

Marvell is regarded as one of the finest Metaphysical poets. His brilliant use of conceits and luxuriant imagery is ever-present in this selection which includes much of his lyrical poetry together with some of his more political works. Such famous poems as *To his Coy Mistress*, the Mower poems, and *On a Drop of Dew* can all be found, with informative notes and introduction by Frank Kermode and Keith Walker.